The Feather Bender's
ADVANCED
FLYTYING
TECHNIQUES

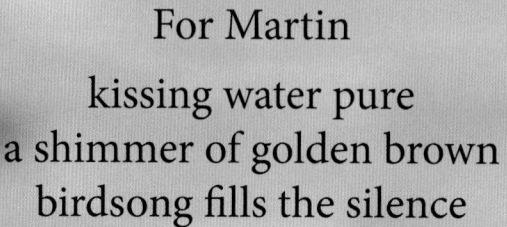

For Martin

kissing water pure
a shimmer of golden brown
birdsong fills the silence

L&C

B

The Feather Bender's
ADVANCED
FLYTYING
TECHNIQUES

BARRY ORD CLARKE

Merlin Unwin Books

First published in Great Britain by Merlin Unwin Books Ltd 2024

Text © Barry Ord Clarke 2024
Photographs © Barry Ord Clarke 2024   www.thefeatherbender.com

All rights reserved, including the right to reproduce this book or portions thereof in any form or by any means, electronic or mechanical, including photocopying, recording, or by any information storage and retrieval system, without permission in writing from the publisher:

Merlin Unwin Books Ltd
Palmers House
7 Corve Street
Ludlow
Shropshire SY8 1DB
UK

www.merlinunwin.co.uk

The author asserts his moral right to be identified with this work.
ISBN 978 1 913159 77 1
Typeset in 12 point Minion by Joanne Dovey, Merlin Unwin Books
Printed by Promat

CONTENTS

Foreword by Hans van Klinken	7
How to use this book	8
Hook conversion chart	10
Proportions	13
Flies and buoyancy	17
CDC	17
Gallo de León	17
Flor de Escoba	18
Flytyers' wax	18
Varnishing heads	22

THE FLIES

1	Ubiquitous Nymph	27
2	Danica Mayfly Nymph	35
3	Moose Mane Nymph	43
4	Clarke's All-Purpose Emerger	51
5	Midge Emerger	57
6	Emergent Sparkle Pupa	63
7	Hare's Ear Soft Hackle	71
8	All Fur Wet Fly	79
9	Hare's Ear Parachute	87
10	Anorexic Mayfly	95
11	CDC Mayfly Dun	103
12	Mallard Slip Wings	111
13	Comparadun	119
14	Clarke's Caddis	125
15	Diving Caddis	133
16	Dyret or The Animal	141
17	Welshman's Button	149
18	Fluttering Caddis	157
19	Red Panama	167
20	Giant Stonefly	175
21	Braided Blue Damsel	185
22	The Worm	193
23	Float Foam Ant	205
24	Madam X	213
25	Foam Cylinder Cranefly	221
26	Phantom Zonker	229
27	Gummi Grub	237
28	Silicone Fry	245

FOREWORD
by Hans van Klinken

I have no idea where I met Barry for the first time because I saw him at so many flytying shows and flyfishing events where we were tying flies together. Everybody who ties flies at fairs knows that it is very hard to talk with other flytyers during the events. For a serious talk, we had to wait till dinner afterwards because only then was it possible for us to have a nice chat and share our favourite tales. Barry and I always get on with each other very well, probably because we have so much in common. We highly respect each other; Barry is a true gentleman, always kind and polite to everybody and always willing to help people. He is also lovely with kids who try to learn all about flytying.

Barry and I write for flyfishing magazines all over the world, often the same ones. This is why I know so much about Barry's fantastic photographs and beautiful artistry at the vice, and how I was able to follow the development of his flytying skills over so many years.

Barry's work is the perfect welcome into the world of flytying whether via his amazing website, The Feather Bender, or one of his many publications. I have a nice set of his flytying booklets that I frequently use during my kid classes in Bosnia, Scandinavia or Canada. Believe me, kids don't mind the language barrier; and the tying steps as Barry photographs them in his books, articles and booklets are out of this world.

Each individual tying step is in perfect harmony with his threads and feathers. The more tying steps, the clearer the tying process and the quicker people will understand and that's what I like for my own classes, workshops and publications as well. My close flytying friend Leon Links and I also agree about this. I favour creating around 20 tying steps to explain my flies and dressings and Barry often goes even further. As a fanatical photographer myself (Barry and I even shoot with exactly the same cameras) I appreciate just how much work it takes to show each pattern in perfect synchronisation within the tying process. I love clear tying steps, in good composition, shot with high resolution cameras – so Barry is a photographer very close to my heart.

Of course, I love to present my own patterns but often for kids my flies are too complicated and that's why I choose teaching them with patterns as described in Barry's books. Flytyers are often seen by outsiders as crazy people but most of them are moved to share the art with others.

Sadly I haven't fished with Barry yet but I know he is an incredible flyfisher. I not only hear that from those who have fished with him, but I can also tell from the ways in which he explains the tying steps and material choices so clearly and in such detail. He is a big nature lover and I well remember him showing me a herd of deer standing in his garden during one of our recent video calls.

When Barry asked my permission several years ago to do a video about tying a Klinkhåmer Special, I was happy to agree as I knew he would tie my Klink even neater than I would tie it myself. He is an amazing flytyer, a perfectionist on a par with my late friend Oliver Edwards. Barry has produced books for beginners, mid-level and advanced flytyers. His idea to perfectly echo and link the tying steps in his book with the tying videos on his website is absolutely unique.

Without doubt I can confirm that you will learn a lot from this book in which he shares so many useful tricks and tips, in words and pictures, all mirrored in his masterclass videos on his YouTube channel, The Feather Bender.

Barry Ord Clarke, I thank you for all the efforts that you have made, helping this and future generations to become better flytyers.

With best wishes
Hans van Klinken, 2024

HOW TO USE THIS BOOK

This book, the second in my flytying techniques series, has been broken down into chapters and categories. This gives the reader the opportunity to locate the technique or pattern they are looking for with ease. I have carefully chosen each of the patterns, not only to best illustrate many different, slightly more advanced tying techniques, but they also represent some of the very best fishing flies you could have in your fly box.

The index at the rear of the book will tell you where to find particular patterns and techniques. When you have located the chapter for the desired technique, each pattern is listed with a recipe, with recommended hook style, size and materials. These are listed in the order that I use them in the book's step-by-step images and the video. This will help you plan tying each pattern and select and prepare your materials beforehand.

I also cover some of the materials used in this book. I suggest that you use this information, not only to initially familiarise yourself with their uses and applications, but also as a 'go-to' reference when you use one of the materials in a new pattern. Within the patterns throughout the book, I have also listed what to look for when buying these materials: quality, colour, size, variants... This will provide you with essential knowledge that will help you understand each material, its qualities, uses and applications.

I recommend that you start by downloading a QR code app for your mobile or tablet device, then scanning the QR code, or key in the link provided and watch the video of me demonstrating a technique or tying the pattern. Here I describe any special procedure or technique, so you can see, first-hand, how to execute it before you start tying. You can then return to the book and follow the step-by-step instruction to tie at your own speed and leisure. If you subscribe to my YouTube channel, you will be informed whenever I publish a new pattern or technique.

If you are having difficulties with a technique or pattern, you can send me a message via the comments on the video of the pattern in question and I'll get back to you, hopefully with some help and a solution.

Good luck and tight lines.

Barry Ord Clarke, Skien, Norway, 2024

The Feather Bender flies on YouTube

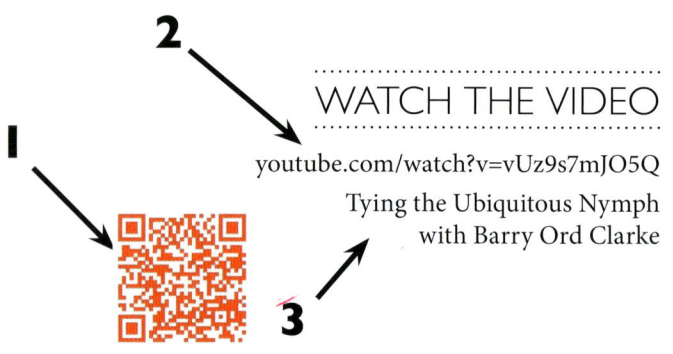

WATCH THE VIDEO

youtube.com/watch?v=vUz9s7mJO5Q

Tying the Ubiquitous Nymph with Barry Ord Clarke

To watch the YouTube videos of the tying sequences in this book you can follow any of these options:

1. To use the QR code: open the camera on your smartphone or chosen device. Hold the camera over the red QR code and your web browser will pop up automatically, leading you to the YouTube video of Barry tying that fly.

2. Or key in to your browser the URL (YouTube link) as shown in the book immediately beneath the 'Watch the Video' heading.

3. Or type in to your web browser the full fly title as it appears at the start of each step-by-step tying section of the book. The video will come up.

Anatomy of the Hook

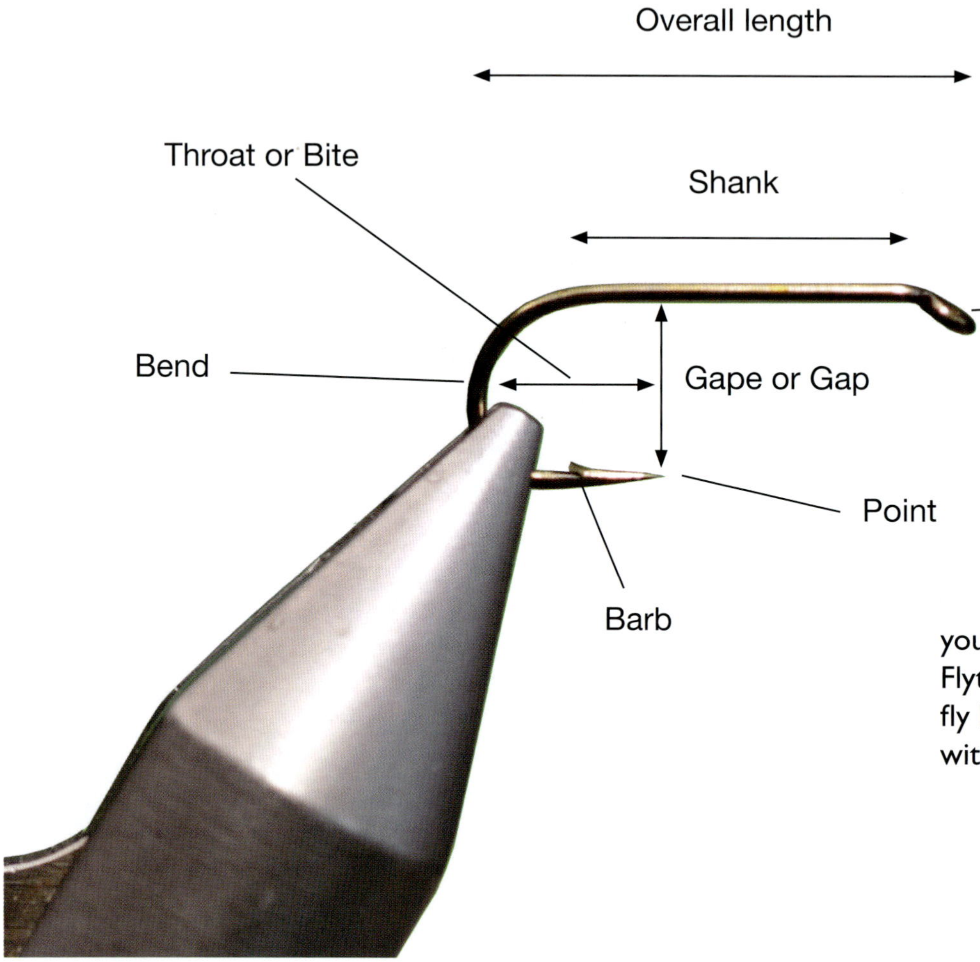

Overall length

Throat or Bite

Shank

Bend

Gape or Gap

Eye

Point

Barb

youtu.be/UUHSnvn3peE
Flytying for Beginners
fly hook anatomy
with Barry Ord Clarke

HOOKS

A hook's size, shape and weight reflects the insect or animal body size and shape it is going to imitate; and how it will sit on the water surface or swim in it!

Dry fly hooks

Dry flies float on the surface of the water, so dry fly hooks are normally fine diameter, made from standard (S) or fine or even extra fine (XF) wire so that there is minimum weight in the hook, making the fly float better.

Wet fly and Nymph hooks

Wet fly (top right) and nymph hooks (bottom right) are both normally made with a heavier diameter wire (XH = extra heavy) to give the hook extra weight, in order to make it sink. Nymph hooks are normally longer than wet fly hooks (2XL or 3XL) to imitate the long, slender bodies of many naturals.

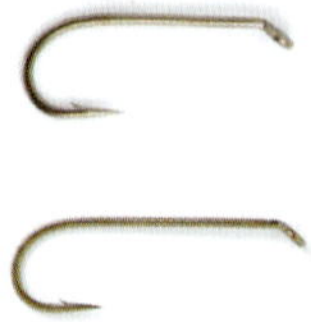

Emerger and Grub hooks

These hooks normally have more bend than the straight hook shank (C = curved). Emerger hooks (top right) are designed to imitate hatching insects that are hanging in the surface film of water. The curved hooks help the flytyer to imitate this stage, with the rear part of the body of the insect submerged and the thorax and wing case above the surface. One of the most successful emerger flies is the Klinkhåmer. Grub hooks (bottom right) have no straight shaft and are totally curved.

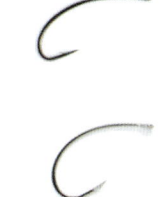

Streamer hooks

Because almost all streamer patterns are tied to imitate bait fish, the hooks used for streamers tend to reflect the natural body shape of a small fish in various sizes. Most streamer hooks are made of standard (S) or heavy diameter wire (XH = extra heavy) and come in various shank lengths (3XL, 4XL).

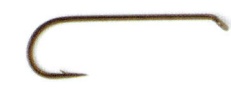

Hook size

Hook sizes cause confusion for most flytyers, let alone beginners. The number on a hook generally refers to its size in relation to other hooks in the series, but there is no industry standard and different manufactures have different standards for applying numbers to their own sizes.

The most important thing to remember is that the size number on a hook packet is a 'relative size' not an actual measurement of a hook. The higher the number, the smaller the hook size with #28 a very small hook. The lower the number, the larger the hook size with #1 a very large hook.

Mustad	Tiemco	Partridge	Ahrex
S82	TMC3761	G3AL	FW500/501
R73 9671	TMC5262	D4AF	FW560/561
C49S	TMC2488	K14A	FW510/511 (516/517 for smaller sizes)
C53S	TMC200R	K12ST	FW530/531
R30	TMC100B	L5A	FW502/503
R75	TMC5363	CS5DE-5X	NS118
C67S	TMC2488H	K5AS	FW540/541
R43	TMC5212	H1A	FW538/539
R50	TMC100B	L5A	FW504/505

  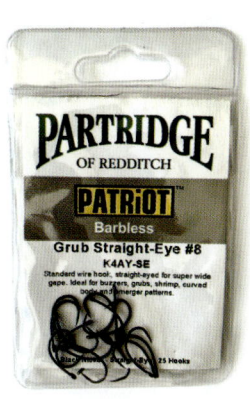 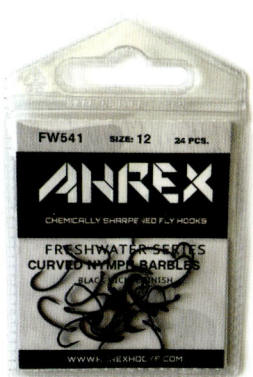

PROPORTIONS

Getting fly proportions right is the Alpha and Omega of a good imitative fly pattern that will deceive trout. Although there are some standard proportions of various styles of fly, they are not set in stone. Many professional flytyers have their own slightly differing styles of signature proportions that can be recognisable with a trained eye.

Tail and wing length, hackle size, position of the thorax and length of abdomen, are all too often overlooked for the sake of technique, and establishing an eye for gauging proportions and achieving consistency takes years of practise and repetitive tying. The very best commercial flytyers are those who consistently produce identical flies.

Here are a few pointers that should help provide a sound foundation for improved proportions and consistency.

Standard Dry Fly

Here the attention to correct proportions and materials used are more critical than on any other style of fly. The primary function of a dry fly is to float and the secondary, to give the trout a deceptive representation, that looks and behaves like a natural. This is best achieved by employing the correct hook size and materials in appropriate amounts and proportions.

Proportions for standard dry fly
Wing height = hook shank length
Tail length = hook shank length
Body length = three-quarters hook shank length
Hackle length = three-quarters length of wing = 1-1.5 hook gape

You can start considering proportions as early as when you attach your tying thread to the hook shank, by using it as a measuring gauge and marking positions along the shank with it as you go. One of the most common faults for many flytyers is crowding the hook eye, or not having enough space left between the hook eye and the thorax to finish the fly cleanly. All this is result of earlier incorrect proportions. This can be somewhat resolved by starting your tying thread 2-3 mm behind the hook eye and adopting this as a finishing point.

Similarly, continue using touching turns of tying thread for the foundation along the hook shank, leaving small open spaces to mark relevant positions for the wing, thorax, abdomen… If further help is required, you can also exercise the use of a bare hook and use this, as you go, as an indicator for proportions.

All these proportions employ a classic standard-sized dry fly hook, such as the Mustad 94840/Mustad Heritage R 50AP. If shorter or longer shank hooks are being used, the proportions should be modified accordingly.

I can also strongly recommend having a collection of well-tied reference patterns in the sizes you use and tie. These can be purchased individually over time, from fly stores and shows. These will be an invaluable resource when referring to correct materials, amounts and proportions.

In addition to the standard proportions of flies there are also the flytyer's personal preferences for under- or over-dressed flies, depending on the pattern or style of fly.

Parachute hackle flies

Parachute-style patterns, be they a standard dry fly style or a Klinkhåmer emerger style, have the hackle wound above the thorax on a horizontal plane. The most defining measurements/proportions are the position and length of the post and hackle size. Both are relative to each other.

On both styles of fly, once the position of the thorax has been established on the hook shank, the parachute post can be located in the centre of this for emerger/Klinkhåmer patterns, or slightly off-centre to the rear of it, for a standard dry fly.

The post or wing, as it may also be called on a parachute pattern, has several functions, both practical and visual. First and foremost, the parachute post is an anchored footing for attaching, wrapping and securing the hackle. But depending on what insect or what life cycle stage the pattern represents, the post can also be used as an up-raised wing, a sight indicator, or both. On some patterns, the post will also function as a sail helping presentation and keeping the fly on an even keel.

Originally the parachute hackle technique was designed to bring the thorax and body of the fly in closer contact with the water, to aid presentation and deliver a better, more lifelike footprint for the trout.

These early parachute flies used white calf tail or calf body hair for the post/wing. Today we have a vast choice of materials that can be employed. Among the more common are various types of polypropylene yarn. But deer hair, CDC and assorted foams are also common in many patterns.

If you are tying mayfly duns or up-wing emergers, the post should be of a size to represent the raised upright wings of a mayfly. If this is the case, in theory, these wings should be long enough for the trout to see from their refracted sub-surface viewing angle. Some flytyers consider these upright wings of great importance, believing they bestow a feeding trout with a sense of urgency, indicating that the insect is about to ascend and thus be taken off the menu!

Bear in mind when trimming your post/wing that it can always be made shorter, but can't be made longer, once cut. Take your time when cutting your post to size, it's better to trim it a little at a time and get it right than too short.

When tying parachute patterns to represent adult or emerging insects that don't have up-raised wings, like caddis and midge patterns, a post of particular length is not required to represent wings. Here the post's main function is as an anchor for the hackle, and it can also double as a sight indicator making the fly easier for the eye to locate at distance. If this is the case the post is trimmed short, so as not to be visible to the trout but still visible to the angler. If your eyesight, like mine, is failing with age, the use of Hi-Viz polypropylene yarns as post material can help significantly.

Hackle amount and size, for parachute patterns, leans very much toward personal preference. Some anglers like a large sparse hackle, underdressed, with only a couple of turns around the post. Others prefer a shorter denser hackle wound from higher up the post. I personally am a believer in that 'less is more' which has for me resulted in more fish to the net. But again you can follow the standard dry fly hackle rule for the parachute of 1-1.5 length of the wing/post.

Nymphs

Because the naturals vary greatly in size and shape, thick and thin, short and flat, long and slender, it would be futile to even try and generalise a scale for natural proportions. But if we take a generic nymph as a standard model, let's say a pheasant tail nymph, we can use this as a starting point and build on it.

Some flytyers are drawn towards simplicity when designing nymphs, using basic patterns that centre around size and colour only. Others go all out for ultra-realistic patterns and identical duplication. I personally am of the school of tyers that focus their attention on distinct features, like tails, legs, gills and feelers.

Proportions for a standard nymph

Tail Length = 1/2-2/3 hook shank length

Thorax/wing case = 1/2-1/3 hook shank length

Abdomen Length = 1/2-3/5 hook shank length

Rib = 5-6 even turns of chosen ribbing material

The base of a nymph's tail should be tied in where the hook shank ends and the hook bend begins. This point is normally located directly over the hook barb.

A nymph's abdomen is tapered from thin at the tail base and increasing in thickness to where it becomes the thorax, rather like an icecream cone. The rib that covers only the abdomen should be no more than 5-6 evenly spaced turns of your chosen ribbing material, be it wire, tinsel, floss etc. The rib not only assumes the role of the natural body segmentation of a nymph, but also strengthens and prolongs the life of the more fragile dubbed abdomen.

On my nymphs, from where the abdomen ends, I like the thorax to reverse taper from slightly thicker than the abdomen and ending tapered smaller behind the hook eye.

I feel that emphasising features on a pattern like gills and legs make the nymph more easily recognisable for the trout and bestows a more lifelike animation to my sub-aquatic creations.

When explaining this style of tying to other flytyers, I use the sombrero theory.

If you have a drawing of a matchstick man, can you tell me what nationality he is? You can see that he is a matchstick man, that's clear, but determining what nationality he is, is impossible.

But if I use the same matchstick man, and now draw a sombrero on him, he is instantly recognisable as Mexican! The theory is that emphasising one or more key features on a basic nymph can make a significant difference.

FLIES AND BUOYANCY

CDC

CDC is short for Cul de Canard or Croupion de Canard, (translated as end of the duck) and was first used as a flytying material in the 1920s in Switzerland. In more recent years, amongst others the Swiss perfectionist Marc Petitjean has been responsible for popularising the use of this material.

All birds have a preening gland (uropygial gland) and surrounding feathers, but the best for flytying comes from ducks. The feathers are located around the gland that produces a waxy preening oil secretion. This highly water-repellant oil is collected on these small feathers, and it is here the bird obtains the oil with its bill, to dress the rest of its plumage. Without this oil the bird would become waterlogged and drown.

The small barbule fibres on a CDC feather snag tiny bubbles of air that work not only on dry fly and emerger patterns but also nymphs. Besides its excellent floating properties, CDC is also extremely aqua-dynamic, pulsating with life in the water, and also aero-dynamic. A CDC feather will collapse under air pressure while casting, but as soon as the cast ends, the feather opens and falls perfectly back into its intended shape.

I obtain what CDC I can from my good friend, Mr CDC himself, David Jedlicka from Prague. David painstakingly sorts, grades and sizes each and every single CDC feather by hand. These top quality feathers can be as large as 5 cm long! The very unicorn of CDC! But more commonly in sizes 3.5, 4 and 4.5 cm. If you ever get the chance to get your hands on any of his CDC products, jump on it.

Natural untreated CDC feathers retain the natural oils and features that originally made it so popular through the flies of Charles Bickel and Louis Veya.

Sadly much of the commercial product marketed as CDC today is far from the original. Washing and cleaning with strong detergents and colouring with hot dye baths render much commercial CDC stripped of any of the qualities that it may have had in its natural state. Some suppliers even go so far as to impregnate treated and dyed CDC with a silicone to help it float again.

Gallo de León

The 'Gallo de León' (commonly known as 'Coq de León') that I use, are the real deal, from Javier Escanciano, who continued to breed with the roosters from the President of the Breeders Association in the northeast of Spain, and they shouldn't be confused with anything else.

Flor de Escoba

These hackles are found on only two rare breeds of rooster, 'Gallo Pardo de León' and 'Gallo Indio de León'. The Pardo hackles range in colour from a pale speckled whiteish, golden ginger to a deep reddish brown. The Indio covers a huge range of clear colours from light to dark tones of dun, to subtle shades of steel grey and deep rusty browns.

History tells us that these are without doubt the oldest breed of bird to be bred specifically for flytying. Juan de Bergara, author of 17th century 'Astorga Manuscript' lists over 30 wet and dry fly patterns tied with only silk and 'Gallo de León' hackles, for fishing in the León region. The majority of the fly patterns carefully blend feathers of both breeds with fine coloured silks, trying to imitate the target insects perfectly.

In these long-established Spanish patterns in which they are used, they are not wound around the hook as a traditional hackle. Instead, small bunches of the fibres are tied onto the hook and then splayed with the tying thread to obtain a radial, fan-like hackle.

In recent years, the growing popularity of these hackles amongst flytyers has made them easier to obtain.

Hackles from these beautiful roosters possess rare qualities found in no other feathers. They have brilliant, translucent, almost glass-like luminescent lustre, sometimes verging on metallic. They are blessed with super-fine, long, straight, stiff yet flexible barbs, and a palette of mottled and earthy colours that could only be achieved by Mother Nature and a methodic genetic selection over centuries by their stewards. These special characteristics make these feathers ideal for the finest tails, wings and hackles, not only on dry flies but also wet flies and nymphs.

FLYTYERS' WAX

Today, flytyers' wax is probably the most understated, low-cost, yet invaluable material available to the flytyer.

Although there is a multitude of products that are marketed as 'Flytying Wax', I believe there are only two waxes of significant importance to the flytyer: finger wax (all round) and dubbing wax. If you can find a single workable wax that covers both these aspects, this is all that you will need for all your flytying requirements.

Before the introduction of the bobbin holder and pre-waxed tying thread, flytyers used wax to precondition individual lengths of tying silk as their initial operation before commencing with any tying procedure, so that the amount and adherent qualities of the wax would bind materials in place even when tension was released from the tying thread.

The procedure of applying flytyers' wax to tying thread is a simple and straightforward one that, if practised over time, soon becomes second nature. Commercial flytying finger waxes are generally sold as various forms of cake, and tacky waxes, in lipstick-style dispensers.

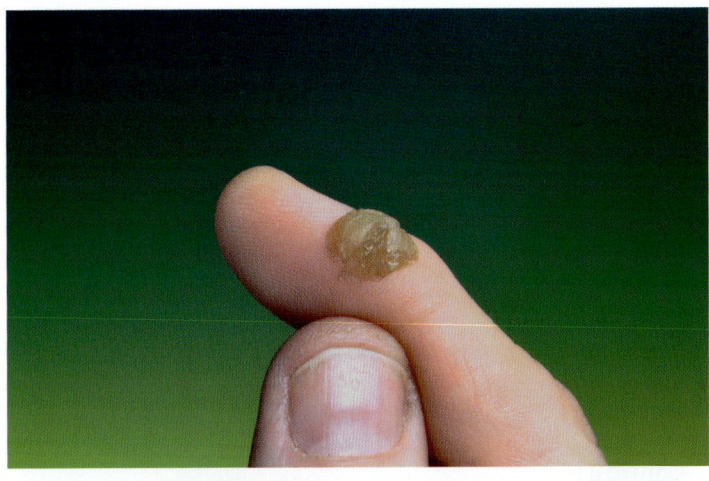

If you have watched any of my flytying videos online, or seen me tie at shows, you will probably know that when tying I keep a small split pea-sized amount of tying wax on the first joint of my right index finger. This positioning of the wax not only keeps it soft and pliable from the natural warmth of my hand but also makes it instantly available for use when required.

To wax the tying thread with this method, I clench the tying thread between the wax on my index finger and my right hand thumb, while I draw downwards vigorously, over 10-15 cm of tying thread, between the hook and the bobbin holder. This creates friction and heat, softening the wax and leaving a residue on the tying thread.

Tyers' wax not only binds materials in place, waterproofs tying thread and stops it from rotting but will also prolong the overall wear and tear of flies generally. If you use modern gel spun tying threads, such as GSP, Dyneema and Nano silks that are smooth and slippery by nature and have little if no stretch at all, the use of flytying wax will facilitate tying procedures that would otherwise prove challenging with gel spun threads.

Tacky dubbing waxes can also prove useful when dubbing some natural and synthetic dubbing materials such as seals' fur and coarse synthetic/natural blends of dubbing. The tacky wax helps the dubbing to adhere to the tying thread if applied a little at a time.

Super sticky waxes that require application with the fingers should be avoided, as these will leave a tacky residue on the hands that will hinder any further handling of materials.

Being a beekeeper myself, I have over the years ventured to synthesise the perfect flytying wax. To say the least, this was a terribly messy procedure and any pans, vessels and utensils used in the operation are rendered useless for anything again, other than the melting and blending of wax!

Taking into consideration that a few small cakes of tyers' wax will last the average flytyer a lifetime and unless you intend to produce flytying wax on a commercial scale, I recommend that you keep things simple and stay with store bought flytyers' wax.

One of the best store bought waxes available is Veniard's prepared flytyers' wax, popularised by Davie McPhail in his YouTube videos. This is a classic cake of all-round tyers' wax that will serve most applications required by the flytyer.

That being said, if you do wish to embark on the journey of making your own wax, here's my formula for an all-purpose wax I have developed over the years and use in all my flytying.

All-purpose flytying wax formula
65% rosin (resin obtained from pine trees)
20% natural beeswax (cleaned)
15% castor oil

- All measurements are in weight. ie; for 100g of finished flytying wax: 65g rosin, 20g beeswax, 15g castor oil.
- Melt the beeswax in an old saucepan, on a low heat while constantly stirring.
- Once the wax is fluid, add the oil and continue stirring.
- When the beeswax and oil are well mixed start to add the rosin while constantly stirring until it acquires a thick clear golden honey consistency.

Once well mixed, I use a small silicone chocolate mould to form the skull cakes. The cakes need time to cool and set before they can be removed from the mould. If you intend making more than one mould of cakes, the wax mixture should be kept warm to enable pouring later. As the wax cools it changes colour from a clear golden honey to a beige blended honey colour.

If you haven't used flytyers' wax, I can highly recommend that you start. Like anything it takes time to adjust to using it, but used correctly it will make your flytying easier, prolong the life of your flies and generally improve your skill as a flytyer.

VARNISHING HEADS

Head cements and varnishes used in flytying have come a long way in recent years. In certain branches of our craft, purists still take the view that glue should never be used in fly-dressing and would refer dismissively to their use, especially to such work that would need liberal application of glues or resins. Today however, cements, adhesives and especially UV-resins have been embraced by the modern flytyer and are here to stay for the foreseeable future.

Apart from those offered specifically for the flytyer through flytying retailers, many flytyers look along the nail varnish shelves in stores and opt for a topcoat of Sally Hansen's Hard as Nails for a super high gloss finish.

Finishing the head of a fly with varnish or head cement sounds straightforward enough, but many a perfectly tied fly falls from grace because of a badly finished head.

Tools

Not all dubbing needles are made equal!

The most essential tool for varnishing heads is your dubbing needle. Albeit the simplest tool on the flytyers' bench, the dubbing needle has many tasks to perform: the application of varnish, picking out fibres from dubbing, splitting tying thread, mixing resins, etc. And there is a huge difference between a well-designed dubbing needle and one that is unworkable.

A good dubbing needle should have a well-formed, easily held, but not cumbersome handle, about the same diameter as a chopstick. The needle itself should be short (4-5cm) and fine with a sharp point. Avoid dubbing needles with long or thick needles – they are hopeless for precise work. You can prove this for yourself: try applying a tiny drop of varnish precisely to a head with a long needle and then with a short one. You will never use a long dubbing needle again!

A long needle will also submerge deep into a small bottle of varnish, drawing much more varnish than required. A short needle, on the other hand, will only reach the surface of the varnish in the bottle, drawing just a small amount of varnish.

The sharp end of a dubbing needle can quickly become covered with a build-up of varnish or resin. This can be scraped away with a sharp blade, but I keep needles clean and sharp with two methods. I have a small plastic canister that I have filled with fine wire wool. All I need to do is push my dubbing needle through the canister top, and down into the wire wool a few times and the needle point is clean. To keep my dubbing needles sharp, I use very fine glass paper to sand them down to a extremely fine point.

Because of the general state of chaos of the tying bench, I like to have several dubbing needles to hand, so time isn't spent trying to looking for them when required.

Technique

Perfect heads on flies begin with tying thread control. For those of you who have read any of my books, watched me at shows or taken a flytying class with me, you will know my thoughts regarding thread control. This I have preached religiously for several decades.

I use thread control throughout all my flytying, but once you have finished the fly and all that remains is the concluding head, thread control is the foundation for excellence. This starts with the last few wraps of tying thread to tidy up the head before you whip-finish.

This is achieved by spinning your bobbin holder anti-clockwise. This gives your tying thread a flat profile, rather like a very fine floss silk. Starting with your tying thread tight into the hook eye, start wrapping rearward, working your way to the back of the head. This 'opening' of the fibres of your tying thread will allow each final wrap of thread to fall perfectly within any previously irregular and uneven wraps of thread, smoothing out the surface.

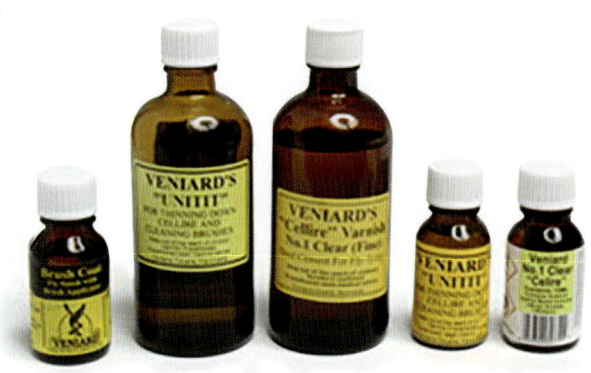

Once you reach the rear of the head, using a whip-finish tool, make a whip-finish or two. The reason I suggest using a whip-finish tool over any other method is that you can place each turn of thread precisely with the tool. This is not possible with your fingers or a half-hitch tool.

Chemistry

Now you can give the head the first coat of varnish. This should be a clear, fine spirit-based varnish that will soak into the tying thread and penetrate the head. My choice is Veniard Number 1 Cellire varnish. This is a clear, low viscosity varnish that dries quickly and, after several coats, results in a hard glossy finish. Most varnishes tend to harden with the passage of time, especially if the lid is left off the bottle too long. Veniard produce a Unitit Thinner for use with their Cellire varnish. This can be used to thin down the varnish to just about any viscosity.

Once you have given the head its initial first coat of low viscosity varnish, it will be absorbed into the tying thread. You should wait a short while before it receives the second.

The second coat will adhere to the surface of the head, filling out any irregularities in the tying thread.

The following coats are the finishing coats and can be given until the desired results are achieved.

Avoid using UV-resin as the only coat given to the head of the fly. Any resin that is absorbed into the tying thread will never cure, as the UV light cannot penetrate into the resin absorbed into the tying thread.

Last but not least. Steady your hands. Applying varnish with a 'free hand' without any support to steady it, is almost impossible. So steady your hands. This is easily achieved by resting your left hand on your vice, and then steadying your right (varnishing hand) with your left hand.

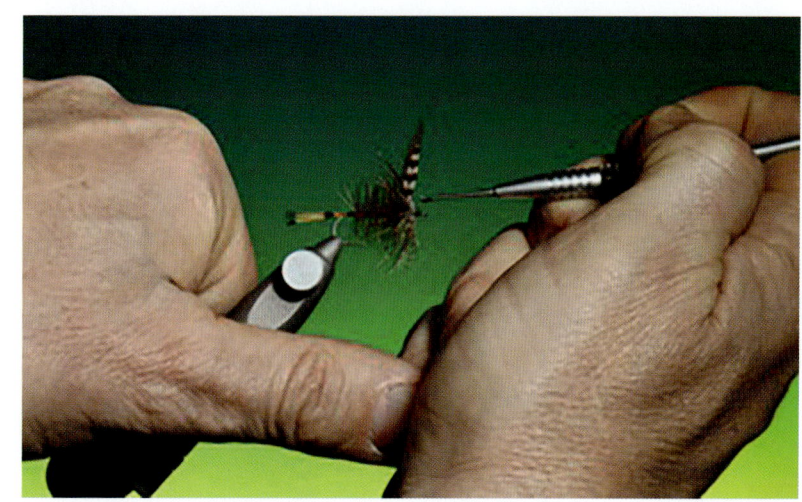

THE FLIES

1

Ubiquitous Nymph

Split mayfly tails • Perfect wing cases • Split pheasant tail legs

A generic nymph for all-round use, anywhere, anytime. This little nymph over the years has worked wonders for me. Rather than targeting any single nymph species, the design tries to offer as many 'nymphy' qualities as possible, as reflected in its name.

The materials used here for the Ubiquitous are the 'Holy Trinity' for trout flies: hare's ear dubbing, pheasant tail and peacock herl. With the addition of the two synthetics, fibbets and oval gold tinsel, we have a classic buggy nymph-style that when all combined give the impression of life.

The hare's ear dubbing is as buggy as dubbing gets, and I like to make a whole load in one sitting. I do this with a fine electric beard trimming razor and shave the ears from a hare's mask. If your beard trimmer has a adjustable trimming length, you can set it to cut just 2-3mm tips of the mask hair, and mix this well with the ear material. This makes a wonderful spiky natural dubbing that has many uses.

I like to fish this pattern two ways. The first is as you would a traditional nymph, on a long fine tippet, which enables it to swim naturally, and I use it to search pocket water or to fish blind on a still water. The second is a more modern technique, the 'Klink and Dink'. This method uses a Klinkhåmer pattern on the point of your tippet with another length of tippet tied to the Klinkhåmer hook bend. Normally this extension tippet is about one metre or longer, depending on how deep it is where you are fishing, with the nymph tied on the end of it. Some flyfishers choose the Klink and Dink extension tippet technique, but take care if using barbless Klinkhåmer hooks! Your extension tippet when tied to the bend of a *barbless* hook can slip off!

Although the latter technique is quick to rig and fishes well, I prefer to use a tiny silver tippet ring, tied in at the very end of the abdomen of the Klinkhåmer, that you attach the tippet extension to. This allows the nymph to drift more naturally, with an unrestricted swimming action, because the tippet isn't fixed, but swings freely on the ring. *See step 28.*

This is a technique for fishing both a dry fly/emerger and a nymph at the same time. It is extremely effective for searching water, when there is no evident feeding activity.

The ring is attached by a short length of braided line before you start to tie the fly. These rings are available in several sizes and are well worth trying. This rig will increase your catches.

TECHNIQUES MASTERED

Split mayfly tails
- An easy technique for splitting two or three fibres for tails with only a short length of tying thread that will keep them in position.

Perfect wing cases
- How to make perfectly shaped wing cases for all nymph types.

Split pheasant tail legs
- How to prepare and execute a pheasant tail fibre bunch for well-balanced nymph legs.

Tying the Ubiquitous Nymph

THE DRESSING

Hook: Mustad Heritage R75 # 10-16
Tying thread: Sheer 14/0 brown
Weight: Lead wire (optional)
Tail: 3 fibbets
Abdomen: Hare's ear dubbing blend
Rib: Oval gold tinsel
Wing case: Pheasant tail fibres
Thorax: Peacock herl
Legs: Pheasant tail fibres

WATCH THE VIDEO

youtube.com/watch?v=vUz9s7mJO5Q
Tying the Ubiquitous Nymph
with Barry Ord Clarke

youtu.be/Mxb7IGeEICs?

youtu.be/tTnpaIhG7Gk?

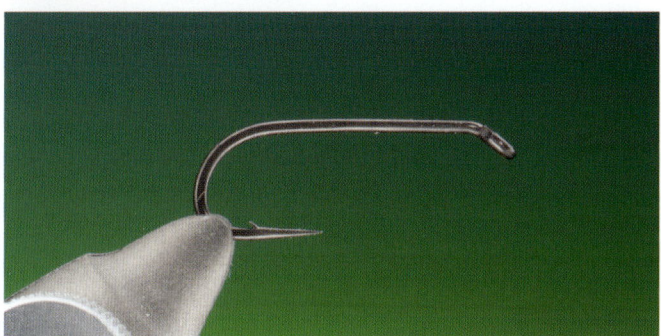

1 Secure your nymph hook in the vice. Make sure that the hook shaft is horizontal. If you have a true rotary vice, centre the hook.

2 If you are going to make a weighted nymph, you will need a short length of fine lead wire.

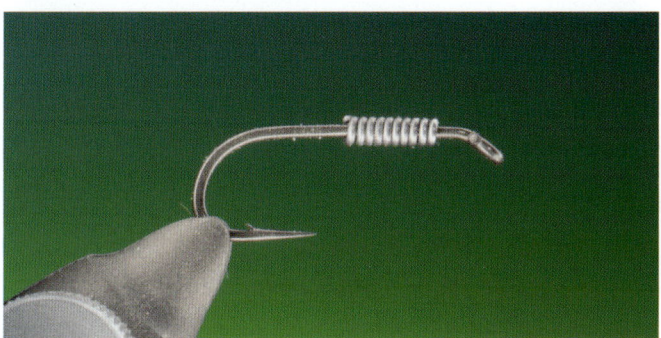

3 Make about 10-12 wraps of lead wire, as shown, a few millimetres behind the hook eye in tight even turns.

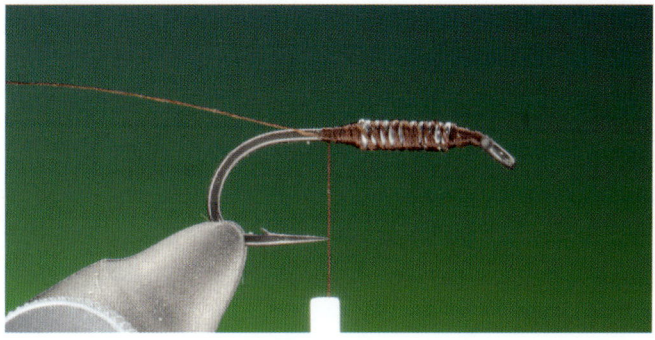

4 Attach your tying thread, forward and tight into the lead wire. Build up a little stopper with tying thread, in front of and behind the lead wire. Remove the short length of surplus tying thread. Don't discard it!

5 You will need the short length of tying thread for the tail later.

6 For the nymph tails, you will need some fibbets. These can be sourced at your tackle shop, or from a fine-bristled paint brush.

7 I use three fibbet tails for this nymph. Take care that you even the tips before tying in.

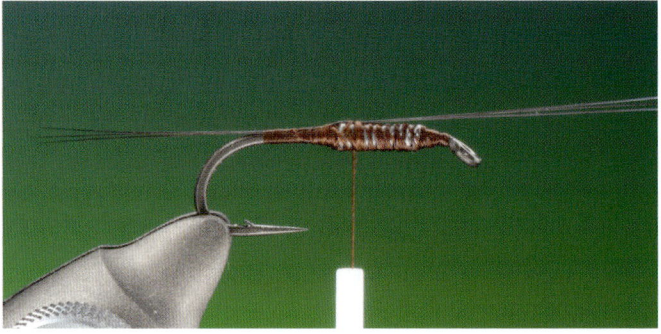

8 The fibbet tails should be tied in on top of the hook shank and be approximately the same length as the hook shank.

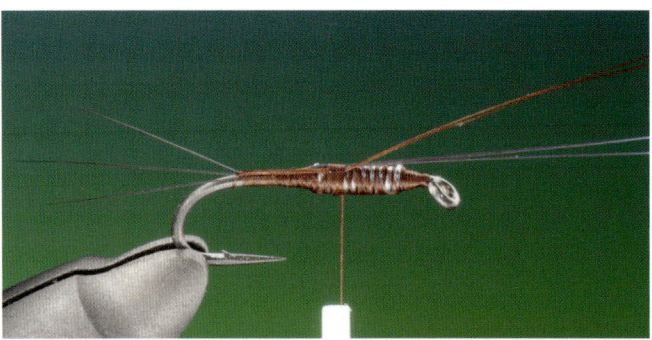

9 Take the short length of surplus tying thread, fold it in two and place it around the hook bend, then use it to split the tails. See video for full instruction.

10 Once the tails are fixed and balanced, remove the excess fibbet material and return your tying thread back to the tail base.

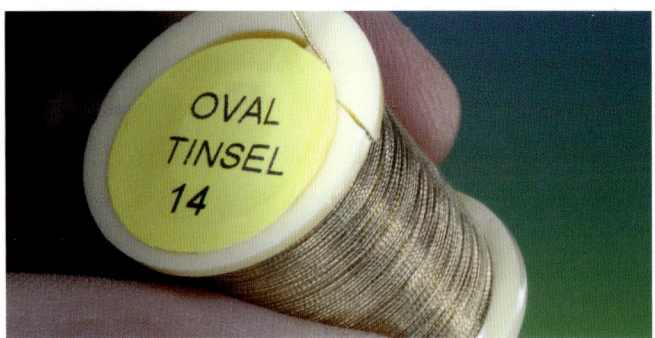

11 You will now need a short length of oval gold tinsel.

12 Rotate your vice and tie in the oval gold tinsel along the hook shank and secure all the way into the tail base as shown.

13 Mix a little short hare's ear dubbing. You can watch my video tutorial for doing this.

14 Wax your tying thread, make a dubbing loop and spin some hare's ear dubbing for the abdomen.

15 Wrap the dubbing forward until you cover the whole abdomen.

16 You can now make 5-6 turns of the oval gold tinsel rib over the abdomen. Tie off.

17 Cut a small bunch of longish pheasant tail fibres, approximately 10-12. Make sure that the tips of the pheasant tail fibres are even.

18 Tie in the pheasant tail fibres as shown, with the tips out over the hook eye.

19 Tie in a single peacock herl where the abdomen meets the thorax.

20 Attach a hackle plier to the peacock herl and wrap in tight touching turns, covering the whole thorax. Tie off and remove the surplus herl.

21 Wax your tying thread and wrap it back to the thorax. Taking hold of the rear pheasant tail fibres, fold them over taking care that they are parallel and secure with a couple of turns of tying thread.

22 Once the wing case is secure, carefully trim away the surplus wing case pheasant tail fibres. Take care that you don't cut away the legs.

23 Split the legs into two equal parts. Take the left half of the legs and tie back to the side as shown, with a couple of wraps of tying thread.

24 Repeat with the legs on the right.

25 Make a couple of whip-finishes and remove your tying thread.

26 You can now give the head a coat of clear varnish. When I tie unweighted nymphs I use black tying thread, so I can distinguish them from the weighted nymphs with a brown head.

27 Bird's eye view of the finished Ubiquitous.

28 The Klinkhåmer fly with the extension tippet ring attached for the Klink and Dink technique.

Danica Mayfly Nymph

Twisted marabou dubbing technique • Pheasant tail body profile • Rear weighted wire ribbing

This is an effective pattern that uses both feather and fibres to create a nymph that pulses and breathes when fished. Both the abdomen and thorax are modelled from twisting a section of a marabou plume into a dubbing rope. This marabou dubbing rope has a very lifelike action and simulates the movement of the natural very nicely with a mass of breathing fibres, the whole length of the body.

Although originally designed as a large Ephemera Danica nymph (Green Drake) these techniques can easily be adapted for alternative nymph species.

Here I show both a subsurface nymph, which can be tied weighted if desired, and a emerger with a foam wing case, so that it hangs effectively in the surface film. With longer tail of pheasant tail fibres tied in over the lower marabou tail, take care that they are long enough to also make the shellback over the whole abdomen. Longer fibres also enable easier handling when tying in.

The 4-5 turns of copper wire rib at the tail base on both patterns not only hold the tail materials in place and add a little bling, but are instrumental in the correct distribution of weight so the patterns fish correctly.

When choosing the first marabou plume needed for the tail and abdomen, look for the tip of one that has very fine tapered unbroken tips. This will produce the best movement when fished. It should also be long enough to wrap the whole body once twisted.

Once the marabou plume is twisted into a dubbing rope and you begin to wrap the abdomen up over the hook shank, your hackle pliers should be rotated, one turn for each wrap around the hook. This will prevent the dubbing rope from unravelling as you progress forward.

TECHNIQUES MASTERED

Twisted marabou dubbing technique
- Selection and preparation of a marabou plume to make a twisted dubbing rope that results in pulsating fibres when fished.

Pheasant tail body profile
- Using a small bunch of pheasant tail fibres to make the tail, shell back and wing case to give the correct nymph profile.

Rear weighted wire ribbing
- A slightly alternative ribbing technique that not only creates the body segmenting but also a little more rear weight for correct presentation.

Tying the Danica Mayfly Nymph

THE DRESSING

Hook: Mustad Heritage C53 # 8
Tying thread: Sheer 14/0 brown
Rib: Flat copper wire
Tail: Cream marabou and pheasant tail
Body: Twisted marabou
Shell back: Pheasant tail
Wing case/legs: Pheasant tail
Thorax: Twisted cream marabou

WATCH THE VIDEO

youtube.com/watch?v=OyFuowmGtTc

Tying the Danica Mayfly Nymph with Barry Ord Clarke

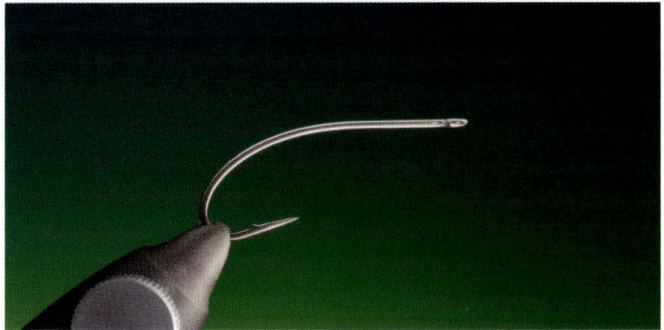

1 Secure your curved nymph hook in the vice. Make sure that the hook shaft is horizontal. If you have a true rotary vice, centre the hook.

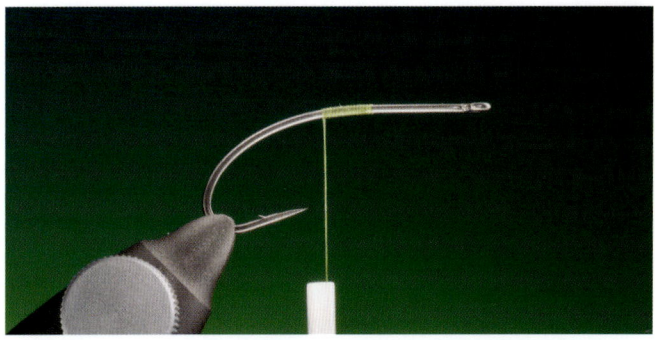

2 Attach your tying thread as shown, central on the hook and run a foundation a short way along the hook shank.

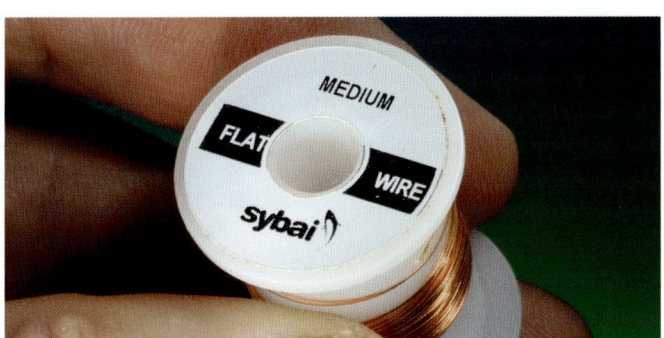

3 You will now need a short length of medium copper wire. For this pattern I prefer to use flat copper wire, but round would be an acceptable substitute.

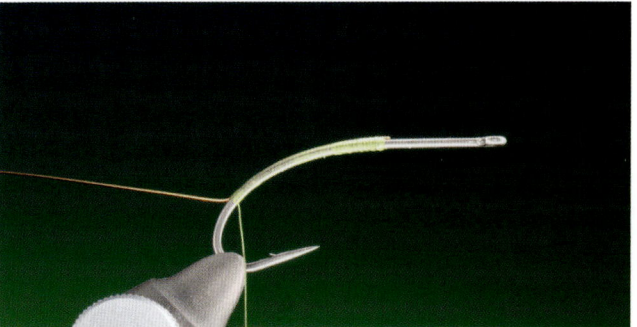

4 Tie in your chosen copper wire, the whole length of the abdomen. This is important for correct weight distribution.

5 When selecting the marabou for the tail and abdomen, look for a plume with nice fine tapered tips, which are even when drawn together.

6 While holding the plume in one hand, pull it through the finger and thumb of the other, drawing all the tips to a small bunch. Spin your bobbin anti-clockwise, so your tying thread attains a flat profile and tie in at the tail base.

7 You will now need a small bunch of long pheasant tail fibres. Tie these in over the marabou tail, approximately twice as long. Once secure, run your tying thread forward.

8 Attach a hackle plier to the end of the marabou plume and twist into a dubbing rope. Take care not to over-twist!

9 Wrap the dubbing rope forward. Your hackle pliers should be rotated, one turn for each wrap around the hook. This will prevent the dubbing rope from unravelling as you progress forward.

10 Once the whole abdomen has been wrapped and tied off, cut away the remaining surplus marabou at the thorax.

11 Take the copper wire rib and make 4-5 close even turns over the tying thread wraps at the tail base.

12 Before you tie down the pheasant tail shellback, run a comb through the fibres so they are straight and parallel. Pull them over the abdomen and make 4-5 turns of rib. Tie off at the thorax.

13 Trim away the surplus pheasant tail fibres.

14 You will now need another slightly larger bunch of pheasant tail fibres. Take care that the tips are even for the nymph legs.

15 Tie in the pheasant tail bunch with the tips out over the hook eye and secure the bunch tight into the abdomen.

16 You will now need some more marabou. This doesn't have to be a plume tip but can be taken as a small section from the stem. This time, tie in by the butt ends not by the tips.

17 Attach your hackle plier and twist into a dubbing rope and wrap the abdomen. Tie off a few millimetres behind the hook eye.

18 Trim off the surplus marabou tips. Now part the pheasant tail into two equal parts and fold backwards along each side of the thorax. Tie down.

19 Here's the side view of the leg position.

20 Take hold of the pheasant tail and fold over the thorax. Run a comb through the fibres to straighten them. Secure behind the hook eye with a couple of turns of tying thread, as close to the thorax as possible.

21 If you are happy with the wing case, trim away the surplus, tie down and build a nice neat head with tying thread.

22 Whip-finish, remove your tying thread and give the head a coat of varnish.

23 Bird's eye view of the finished nymph.

24 For the emerger variant, replace the pheasant tail wing case with a slip of closed cell foam. This will hold the nymph hanging in the surface film.

Moose Mane Nymph

Moose mane quill-style body • Two tone wing case • Stripped hackle nymph legs

From the mane of the North American moose comes the longest hair, especially from an adult bull. This mane is located on the back of the neck of the animal and not from the bell as some flytyers believe. The mane hair ranges from 3" to a huge 9" in length.

The colour is a wonderful salt and pepper mix of white, grey, brown and jet black hairs, but dyed moose mane is also available from some dealers. It has long been used for wrapping quill-style bodies on dry flies and nymphs. The sheer length of the mane hair makes it possible to tie the very largest quill-style bodies on both nymphs and dries.

Tying two hairs in by the tips, one dark, one light in colour, and wrapping them together, creates great segmented bodies. Tying in three hairs of mixed colour can create a deep graduated colour change that has an extremely natural insect body look about it. You can also use a single colour of hair or alternatively, carefully blend the hair colours to match the hatch. I recommend that you experiment with moose mane and the countless possibilities will quickly be revealed.

This pattern shows the flytying potential of moose mane hair. Although I use moose body hair for the tail, shorter moose hairs could also be used for this.

For the best quill body effect, the hairs have to be exactly the same length. You will very quickly discover why if you try to wrap a body with hairs of different lengths. It's also of paramount importance that when you tie in the hairs, the tips are secured tight into each other, without any space between them. Before you begin wrapping the hairs, make sure that they are parallel to each other and not twisted or crossed.

You can then start to wrap the hairs. I find the best way is to use a hand-over-hand wrap. This keeps the hairs parallel. If you have a true rotary vice you could also use this if preferred.

The hackle that I use for the legs on this pattern is put to best use if it has a long sweet spot; this makes the legs the same length. In order to keep all the legs at 90 degrees from the thorax, I also strip the leading edge of the hackle (when wrapped) of all its barbs, before I tie it in.

Although the pattern shown here is weighted, you can tie them without weight as floating nymphs.

Moose mane is significantly more robust than regular deer hair, and you can further prolong the life of moose mane bodies by giving them a coat of varnish. I don't recommend you using UV resin for this as it doesn't adhere well to the waxy coating on moose hair.

TECHNIQUES MASTERED

Moose mane quill-style body
- Moose mane technique for making realistic quill nymph abdomens.

Two tone wing case
- Two tone moose mane wing case.

Stripped hackle nymph legs
- Nymph legs made from a stripped barred ginger hackle.

Tying the Moose Mane Nymph

THE DRESSING

Hook: Mustad Heritage R75 # 8
Tying thread: Sheer 14/0 Brown
Weight: Lead wire
Tail: Moose body hair
Abdomen: Three moose mane hairs
– black, brown, cream
Wing case: Black and Bleached moose mane hair
Thorax: Peacock dub – Olive
Legs: Barred ginger hackle

WATCH THE VIDEO

youtube.com/watch?v=MC7rWuq-EhI&t=16s

Tying the Moose Mane Nymph
with Barry Ord Clarke

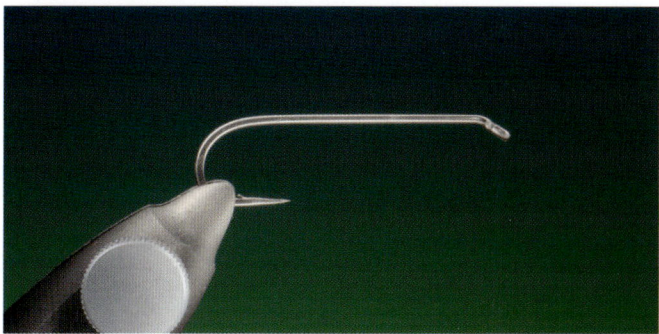

1 Secure your long shank hook in the vice. Make sure that the hook shaft is horizontal. If you have a true rotary vice, centre the hook.

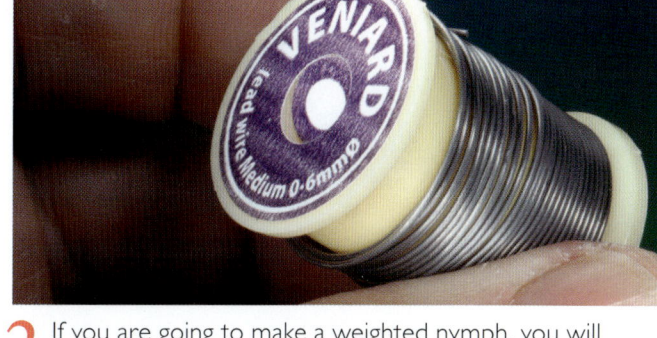

2 If you are going to make a weighted nymph, you will need a short length of medium lead wire.

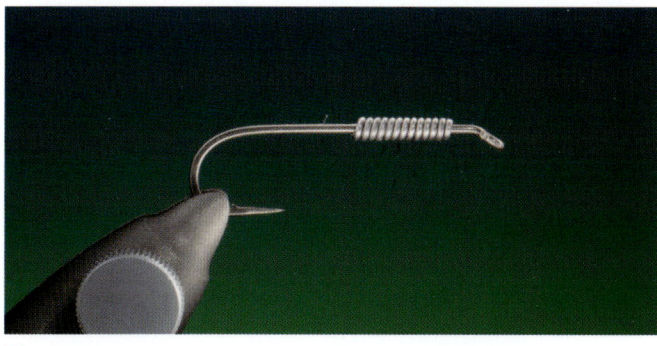

3 Make about 10-12 wraps of lead wire, as shown, a few millimetres behind the hook eye in tight even turns.

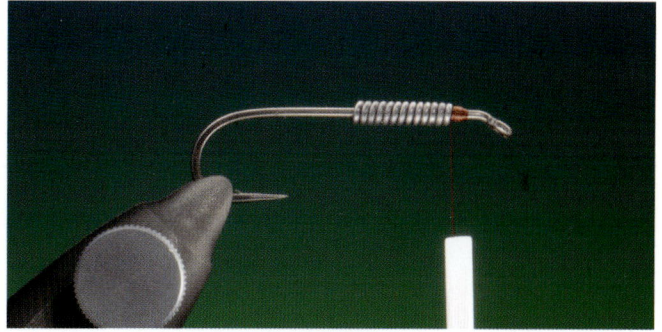

4 Attach your tying thread, forward and tight into the lead wire. Build up a little stopper with tying thread.

45

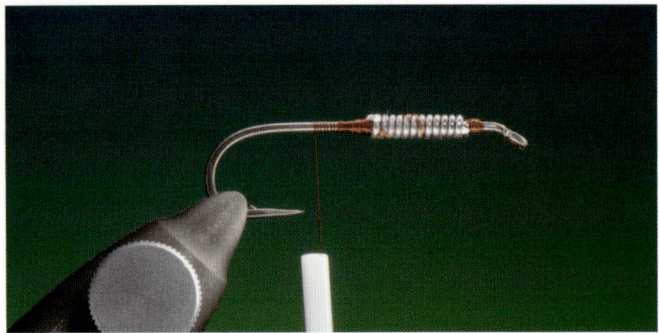

5 Run your tying thread over the lead wire and make another stopper at the end of the lead wire.

6 Select a small bunch of dark moose body hair. Take care that the hairs have fine, unbroken tapered tips.

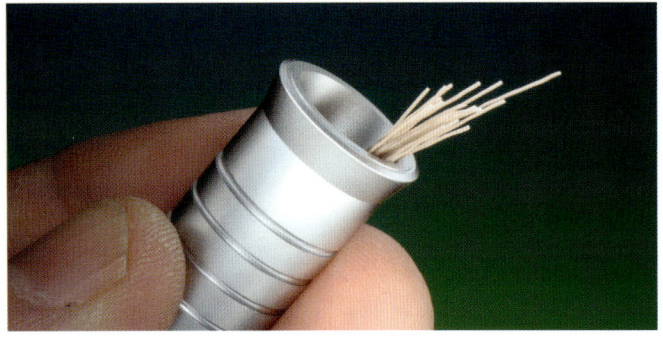

7 Before you tie in the moose body hair tail, you will need to even the tips in a hair stacker.

8 The tail should be approximately the same length as the thorax. Tie in as shown, forward to the lead wire.

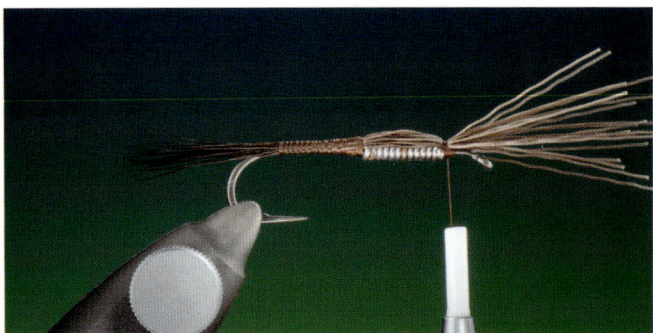

9 Run your tying thread over the lead wire, to the forward stopper. Now fold over the moose hair and secure tight into the lead wire with a few wraps of tying thread.

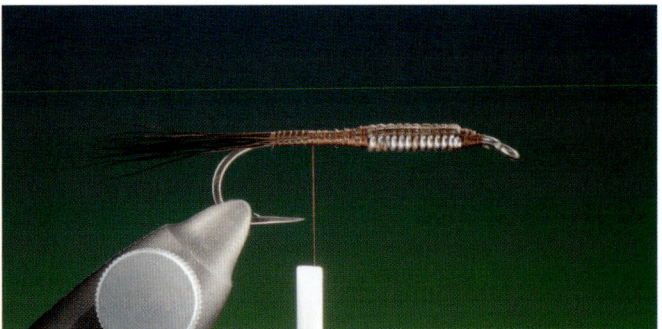

10 Trim away the surplus moose hair and run your tying thread back halfway down the abdomen.

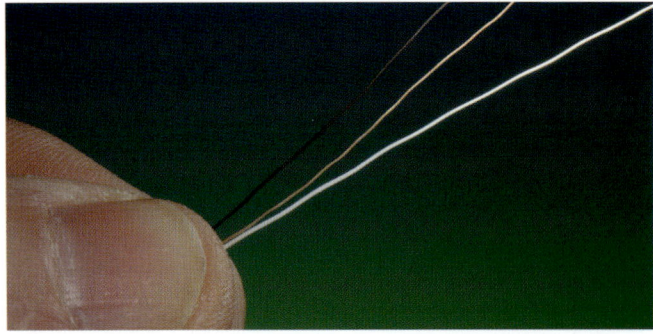

11 Select three long moose mane hairs, one black, one brown and one cream. Align the hair tips. Once the tips are even, trim the hairs to the same length as the shortest hair.

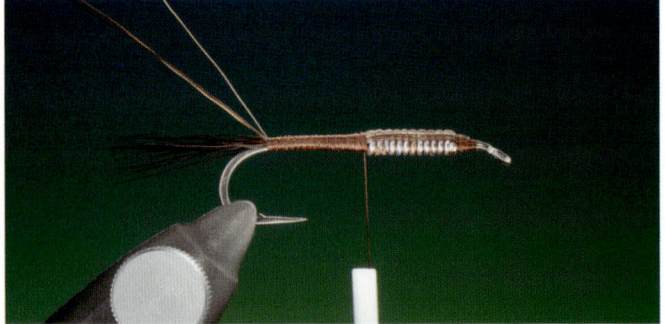

12 Tie in the three hairs by the tips. Secure into the tail base, taking care that all three hairs are tight into each other. You can now use your tying thread to taper the abdomen.

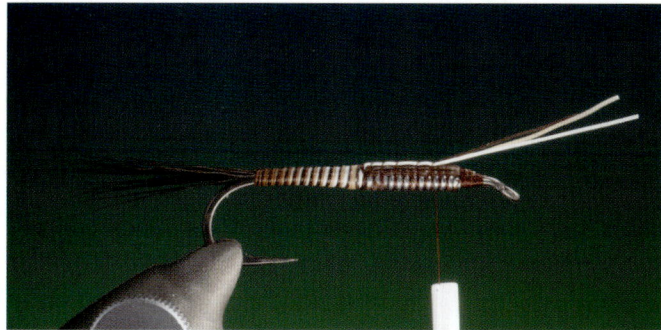

13 Take all three moose mane hairs, taking care that they are parallel and not twisted. Wrap the hairs together in tight touching turns, forward over the abdomen. Tie off at the lead wire as shown.

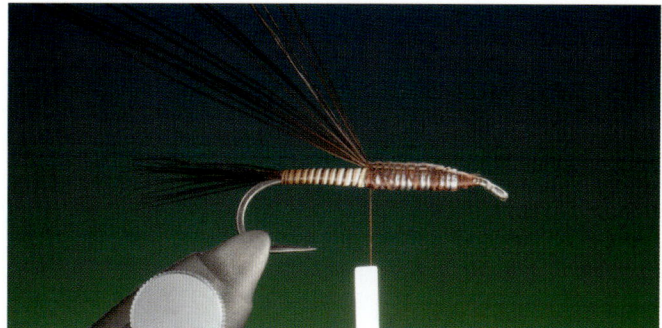

14 You will now need another small bunch of black moose mane hairs. Even up the tips. Tie in the hair by the tips on top of the thorax, keeping the bunch tight together.

15 Cut a slightly larger bunch of bleached moose mane hair, and tie this in on top of the smaller black bunch.

16 Trim away the surplus hair over the hook eye and secure over the abdomen, finishing with your tying thread at the rear of the thorax. Take care not to crowd the hook eye.

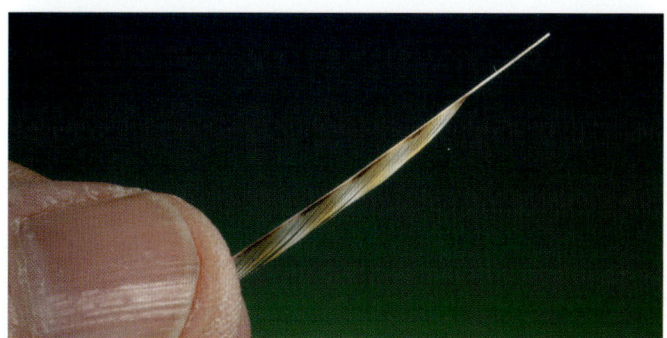

17 Prepare a barred ginger hackle by stripping off all the fibres from the side of the leading edge when wrapped.

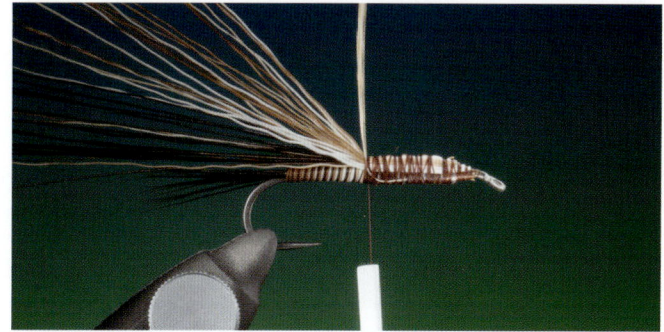

18 Tie in your barred ginger hackle as shown, at the junction of the abdomen and the thorax.

19 You will need some dubbing for the thorax.

20 Dub a broad thorax as shown, taking care to leave enough room behind the hook eye for finishing.

21 Wrap the hackle, palmered style, over the thorax and tie off behind the hook eye. Remove the excess hackle.

22 Now take the bunch of bleached moose mane and fold over the thorax as shown. Take care that the hairs are neat and parallel. Tie down. Note the distance from the hook eye to the tying down point!

23 Take the smaller darker bunch of black moose mane hair and tie this down over the bleached as the central darker portion of the wing case.

24 Trim away the surplus hair, wax your thread, and build up a small neat head with tying thread.

25 Make a couple of whip-finishes, taking care to keep the head small and neat.

26 Finish off by giving the head a couple of coats of clear hard varnish.

27 Ariel view of the finished moose mane nymph, giving a good indication of proportions.

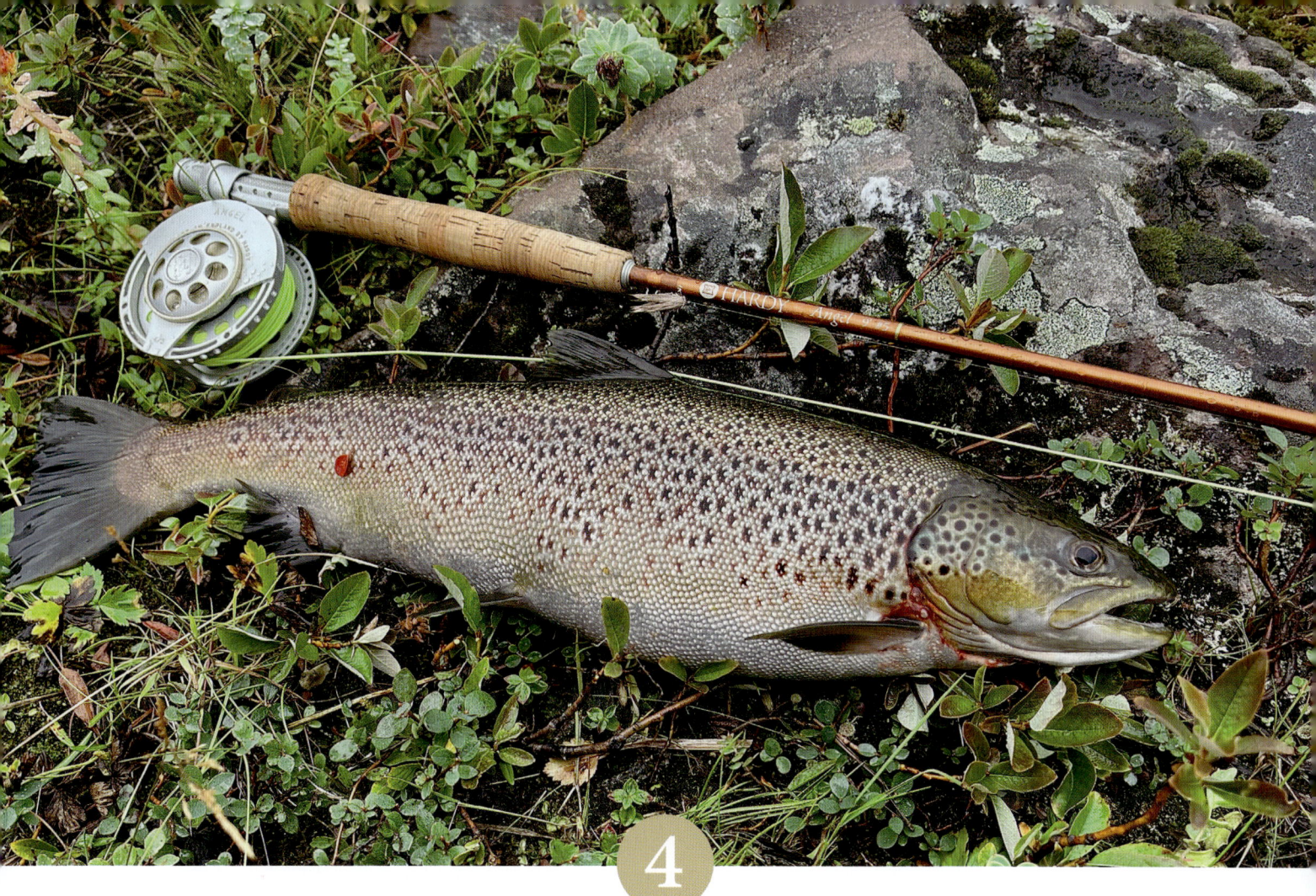

4

Clarke's All-Purpose Emerger

Krystal Flash trailing shuck • Deer hair hackle • Deer hair shell back

This little deer hair bug doesn't exhibit the characteristics typical of many emerger patterns, but don't be fooled…

Emergers represent the vulnerable phase of an aquatic insect's life when they are in the process of transition from a sub-aquatic nymph, larva or pupa, into a winged adult stage of the life cycle.

They are neither dry flies or wet flies, but can be fished as both. They shouldn't fish high and dry like a traditional dry fly, nor should they fish deep like a nymph, but somewhere in between. One of the most decisive factors for successfully fishing emergers, is that the trout finds them in the right place.

This place, the meniscus, is 'unsafe territory' for an emerging aquatic insect! This is the upper tier of the water that maintains surface tension. Insects, especially small ones, often undergo difficulties trying to puncture the meniscus, and have to fight and struggle to break through the surface tension. It is here they are exposed and liable to become trout food.

Whilst engaging in this endeavour, shuck, legs and other body parts merge into a more nondescript mass, rather than anything that resembles an insect.

Once again, I have to reiterate how important your choice of hook and materials is when designing flies.

With most patterns it's relatively straightforward: lightweight hook and materials for a dry fly; and a heavier hook, even with added weight for a subsurface fly.

Here I have used a curved caddis/grub hook with a deep open gape, that will pull the rear of the APE under the surface and the deer hair shell back and hackle that will support the upper half in the surface.

Although I originally designed this pattern as a caddis emerger, I now fish it, as the name suggests, as an All-Purpose emerger. It's worked well under hatches of mayfly, midges and caddis. I tend to rough up the hackle; and dress the head and deer hair hackle only with a drop of liquid floatant. After a cast or two when the rear of the fly has become wet, it assumes the correct posture.

I haven't found any reason for tying it in a range of colours, as this one has proven successful. I have dabbled with one tied in all black, for night fishing. The deer hair used should be buoyant hair from a late season deer. It's also worth using deer hair that has nice fine tapered barred tips.

TECHNIQUES MASTERED

Krystal Flash trailing shuck
- Using the trilobal surface of a single strand of Krystal Flash as an effective trailing shuck.

Deer hair hackle
- A technique for making a deer hair front hackle from a small bunch of stacked deer hair.

Deer hair shell back
- Utilising the remaining deer hair from the hackle bunch as a Humpy-style shell back.

Tying the Clarke's All-Purpose Emerger

THE DRESSING

Hook: Mustad Heritage C49AP # 16-18
Tying thread: Sheer 14/0 tan
Trailing shuck: Pearl Krystal Flash
Hackle: Natural deer hair
Body: Olive brown Super Fine dubbing
Shell back: Natural deer hair

WATCH THE VIDEO

youtube.com/watch?v=jSBsbNIYmr8&t=28s

   Tying the Clarke's All-Purpose Emerger with Barry Ord Clarke

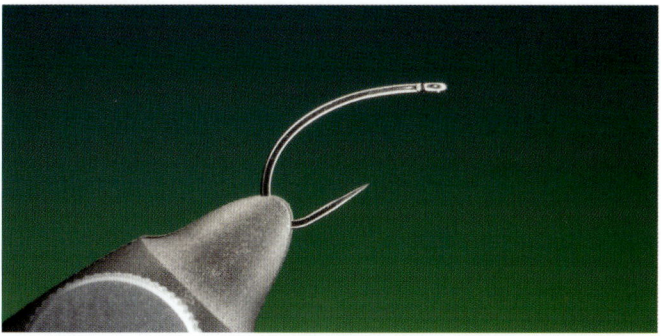

1 Secure your curved grub hook in the vice. Make sure that the hook shaft is horizontal. If you have a true rotary vice, centre the hook.

2 Attach your tying thread as shown, a little behind the hook eye, and run a foundation a short way along the hook shank.

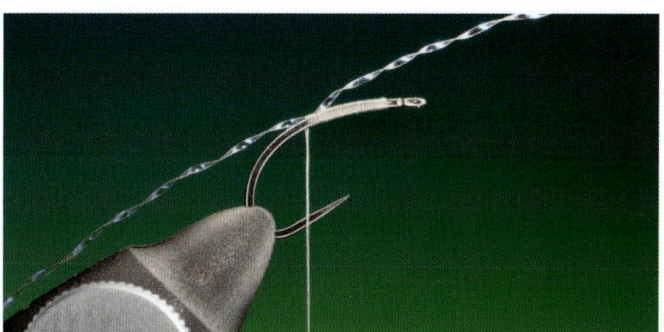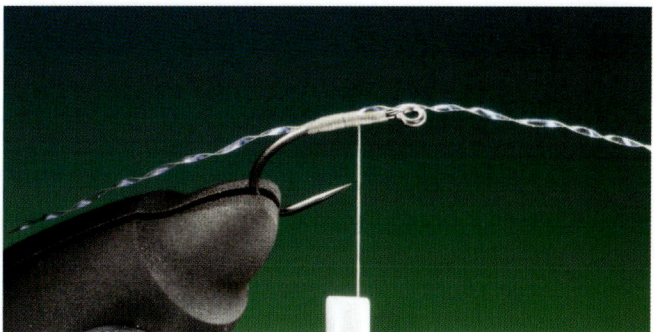

3 You will now need a single strand of pearl Krystal Flash. Attach this as shown with a few wraps of tying thread.

4 If the Krystal Flash strand is lying correctly, wrap forward with your tying thread, securing the strand.

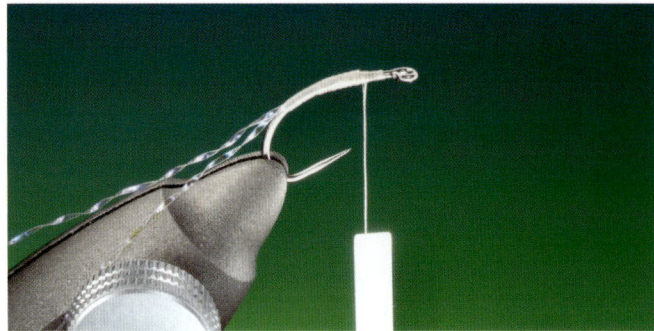

5 Fold the forward Krystal Flash strand rearward and secure down into the hook bend as shown. Finish with your tying thread forward, 3-4mm behind the hook eye.

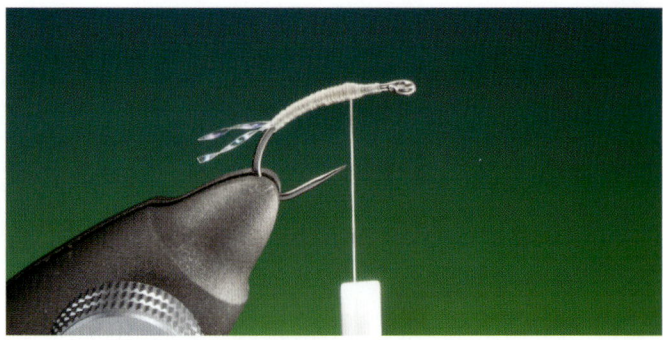

6 Trim off the Krystal Flash trailing shuck so they are just a few millimetres long.

7 Cut and clean a small bunch of deer hair, by removing with a comb all the underfur and the shorter hairs. Stack the bunch in a hair stacker to even up the tips.

8 Tie in the bunch with the tips out over the hook eye. Then secure the hair bunch with a few tight wraps of tying thread, a little way back along the hook shank.

9 Spin a little Super Fine dubbing onto your tying thread, into a tight dubbing rope. Now wrap the whole body, up to the deer hair, with an increasing taper.

10 Once you have a nice dense body, as shown, tie off the dubbing at the deer hair hackle.

11 Spin a little more dubbing tightly onto your tying thread. With your thumb nail, push back the deer hair tips, and wrap the dubbing tight into the deer hair, to support it.

12 Now using a pair of scissors divide the deer hair in two equal parts, and press down on each side of the hook shank.

13 The deer hair should now be divided and pressed out to the sides.

14 Wax your tying thread. Take hold of the deer hair bunch, keeping the hairs parallel, you can run a comb through them so they are straight. Pull the hair tight over the body and fix with a couple of tight wraps of tying thread.

15 Flatten your tying thread by spinning it anti-clockwise, and secure the deer hair. Wrap your tying thread forward, behind the hook eye. Now using scissors or a razor blade, trim the deer hair to a small head.

16 Make a couple of whip-finishes, behind the hook eye, and remove your tying thread.

17 The view of the underside of the All-Purpose Emerger.

18 Scruffy legs of the Clarke's All-Purpose Emerger (CAPE).

5

Midge Emerger

Using the hook to create the correct presentation
• Weighting with flat copper wire • CDC breathing gills and wing

Midges are one of the most abundant species of insect groups found in freshwater ecosystems, which makes them one of the most important to the flyfisher. There are literally thousands of different species of biting and non-biting midges in various stages of their life cycle, available to trout throughout the whole year. Midges represent a significant amount of the trout's diet in both still and running water.

This little pattern fills just about all the criteria for a great fishing fly. It's inexpensive, it doesn't require any special materials or techniques, it's quick and simple to tie and last but not least, it catches fish.

Midge larva come in a wide range of sizes and colours including blood red, olive, grey, brown and black. So all you have to do is change your hook size and the colour of your tying thread and you have most emerger situations covered with this one simple pattern.

It's of paramount importance here that you use the correct # 3 XL (extra long) hook for this pattern.

The hook bend, which is left bare, in conjunction with the extra copper wire weight at the rear of the body, function as an anchor, keeping the emerging midge in the correct position, when hanging in the surface film.

Also take note that the hook size used does not reflect the actual dressing size of the midge, which is half the size of the hook. So choose your hook size appropriately.

The CDC breathing gills and shuttlecock-style wing stop the pattern from drowning. However, I like to dress the CDC wing well with floatant. The wing, when dressed, doubles as a sight indicator, being very easy to keep your eyes on, even at distance.

I fish this on a long fine tippet, dead drifting it on rivers and fishing it static or to rises on still water.

TECHNIQUES MASTERED

Using the hook to create the correct presentation
- Using a 3XL dry fly hook so the fly hangs in the surface film creating the correct presentation.

Weighting with flat copper wire
- Creating a flat copper wire butt and rib that not only indicate segmenting but also hold the fly down at the rear.

CDC breathing gills and wing
- How to use CDC to make the breathing gills of the midge and a shuttlecock-style wing for buoyancy.

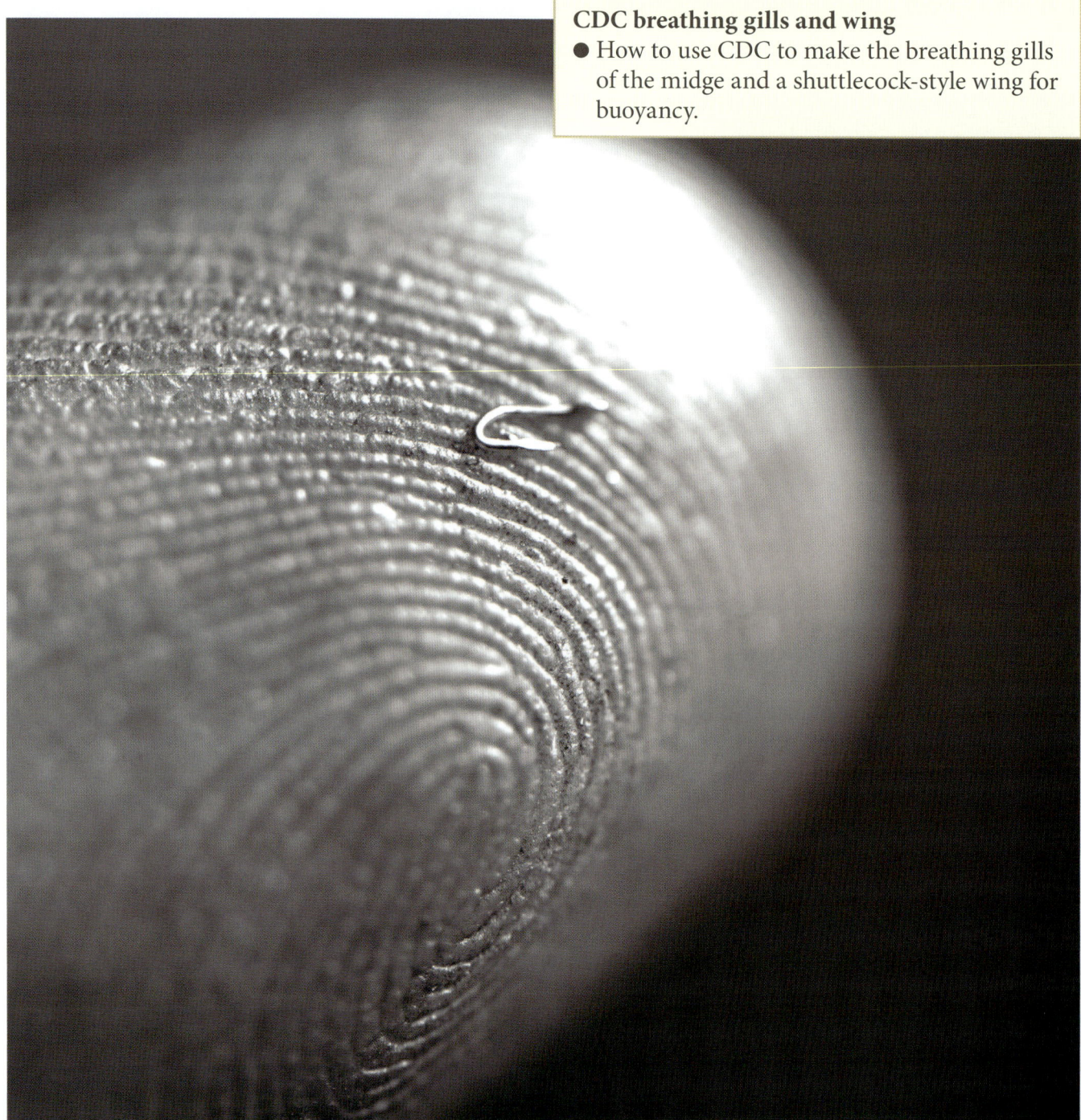

Tying the Midge Emerger

THE DRESSING

Hook: Mustad Heritage R43AP # 14-20
Tying thread: Sheer 14/0 brown
Rib: Flat copper wire
Gills: White CDC fibres
Wing: One or more natural CDC feathers

WATCH THE VIDEO

youtube.com/watch?v=C_b6lQQtGQA&t=307s

 Tying the Midge Emerger with Barry Ord Clarke

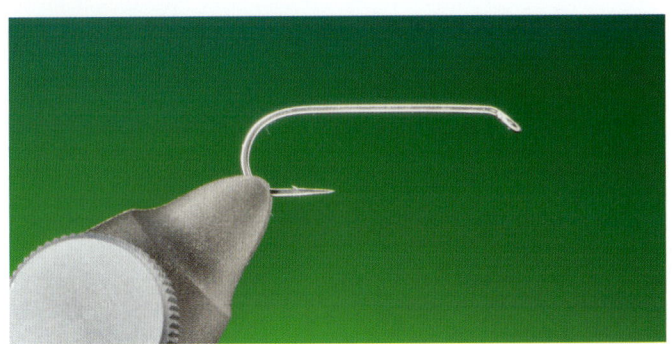

1 Secure your 3 XL dry fly hook in the vice. Make sure that the hook shaft is horizontal. If you have a true rotary vice, centre the hook.

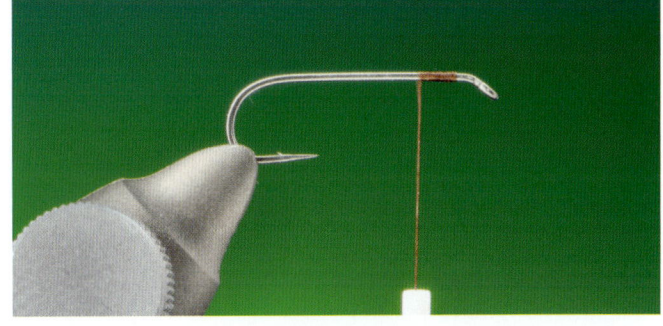

2 Attach your tying thread a little behind the hook eye and cover only a few millimetres of the hook shank.

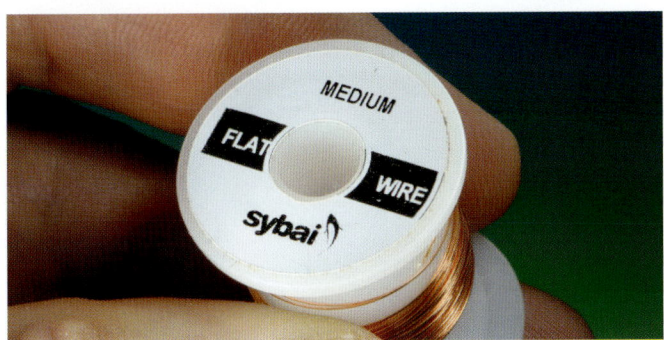

3 I find that a medium flat copper wire works nicely for this pattern, but round copper wire will work just as well, if that's what you have.

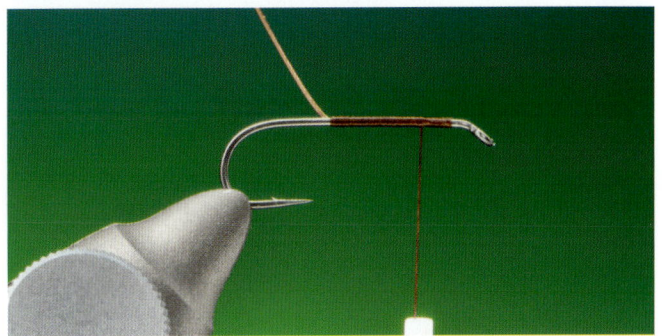

4 Cut a short length of the copper wire and attach it along the rear of the hook shank, little over halfway. Run your tying thread back up the hook shank. Take care with proportions and positions of the materials.

59

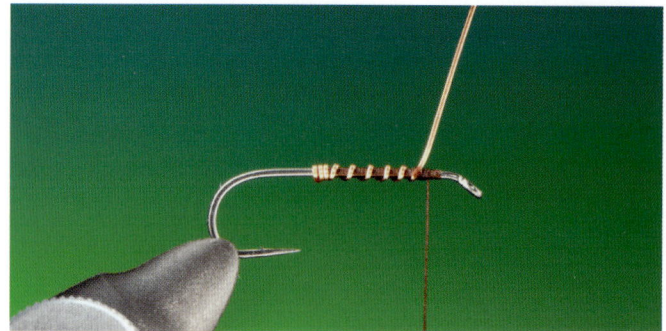

5 Make 3-4 tight wraps of copper wire at the rear of the body, before you wrap forward in 5-6 even open turns. Tie off.

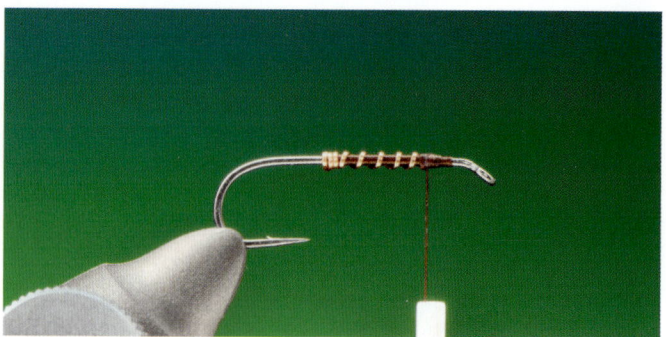

6 Twist off the surplus copper wire and tie down the end.

7 Load a small Petitjean Magic clip with white CDC fibres, for the breathing gills.

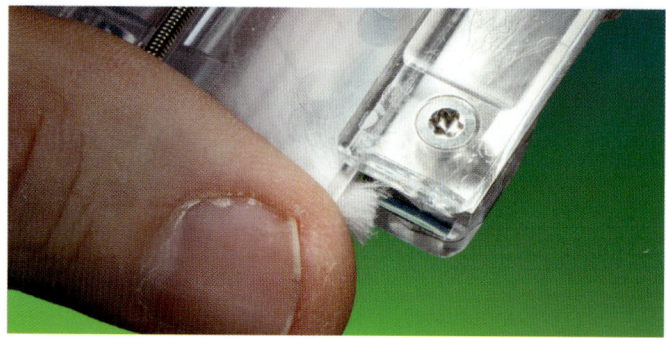

8 Using a Petitjean stacker tool, push all the white CDC fibres together into a small bunch, using your thumb as a stopper, on the edge of the clip jaws.

9 Carefully remove the small bunch of white CDC from the Magic clip. Keep the bunch together.

10 Tie in the bunch with the cut end of the CDC fibres facing backwards, on top of the hook shank.

11 Trim off all the surplus CDC fibres over the hook eye. Then trim down the remaining white CDC bunch into a small tuft, only a couple of millimetres long.

12 Depending on the CDC quality and hook size being used, select one or two natural CDC feathers. Align the tips if using more than one feather.

13 Wax your tying thread. Place the CDC on top of the hook shank and secure in place with a couple of wraps of tying thread. Check that you are happy with the position and wing length.

14 If you are happy, secure with a few more wraps of tying thread. Once fixed, trim off the rear of the CDC wing, as shown, a little shorter than the breathing gills.

15 You can now lift the CDC wing, and make a few turns of tying thread tight in front of it to hold it in position. Support the wing with some wraps of thread around the base.

16 Make a couple of whip-finishes and remove your tying thread.

17 Here you can see the final fishing presentation of the midge emerger, with weight of the rear half of the hook and copper wire submerging the rear abdomen and the CDC shuttlecock wing and gills, keeping it in the surface film.

18 The midge emerger with a buggy hare's ear dubbing thorax for a buggier, higher floating finish.

Emergent Sparkle Pupa

Trailing shuck • Antron yarn overbody • Deer hair wing

This is a variant of one of the many patterns for emerging caddis pupae from Gary LaFontaine's book *Caddisflies*.

The few simple elements and materials used in this pattern's design, are fundamental to its success. The scanty trailing shuck of just a few strands of Antron sparkle yarn are all that are needed to replicate the shedding of the outer pupal skin.

Antron sparkle yarn is a synthetic composite of trilobal fibres, in which each fibre has three distinct sides in cross-section. This enables optimal optical reflective properties. No matter the direction of the light source, it will glisten and mimic life.

This effect is heightened when used to imitate the moulting fluid-filled pupal shuck which surrounds the body. If executed correctly, there will be a small amount of air trapped between the underbody and the Antron shuck, producing the relevant effect.

Unlike the original pattern, here I use a simplified, trouble-free technique for inflating the moulting fluid-filled shuck, that surrounds the body when the caddis fly is propelled to the surface when emerging. LaFontaine recommended that you use a dubbing needle to 'pull' the overbody loose. This technique is cumbersome and somewhat challenging to achieve a balanced result, as you must address the Antron yarn pupal shuck, a little at a time.

Using the technique shown here, you get an instant result and an evenly balanced inflated overbody. This requires you leaving the front third of the hook shank bare, and free from material.

When you have dubbed the underbody, make three or four loose, open spiral turns of tying thread forward to the hook eye. You then fold over the Antron sparkle yarn from the rear of the hook; take care not to bring the trailing shuck yarn with it. Once folded over, make one or two wraps of tying thread, tight into the hook eye, to hold it in position.

Grip the yarn, directly over the hook eye, between the finger and thumb of your right hand and carefully twist from side to side. This is to distribute the Antron yarn around the hook shank and not just above and below.

Once you have done this, make a couple of tighter wraps of tying thread to secure the yarn in position. You then take hold of the two yarn ends and carefully draw them back towards the rear of the hook. If done correctly you will see the pupal shuck magically inflate! See video for full technique.

For the wing, I recommend a fine deer hair with nicely marked tips. You don't want this to flare too much. The wing should be approximately the same length as the hook.

In LaFontaine's original pattern he used marabou for the head. This has over the decades been replaced with dubbing.

The emergent sparkle pupa can be tied in a huge combination of both body and sparkle yarn colours, LaFontaine lists at least 15 of these in his book *Caddisflies*.

TECHNIQUES MASTERED

Trailing shuck
- A simple trailing shuck of a few strands of white Antron yarn representing the loosening sheath as the insect emerges.

Antron yarn overbody
- A special technique for inflating the melting fluid-filled shuck around the underbody as the insect thrusts itself to the surface.

Deer hair wing
- Using a small bunch of fine stiff deer hair for the immature caddisfly wing.

Tying the Emergent Sparkle Pupa

THE DRESSING

Hook: Mustad Heritage R30AP # 12-16
Tying thread: Sheer 14/0 black
Trailing shuck: Cream Antron sparkle yarn
Overbody: Cream or grey Antron sparkle yarn
Underbody: Olive Antron dubbing
Wing: Deer hair
Head: Dark hare's ear Antron dubbing blend

WATCH THE VIDEO

youtube.com/watch?v=XSBIvyCHCbI&t=36s

Tying the Emergent Sparkle Pupa with Barry Ord Clarke

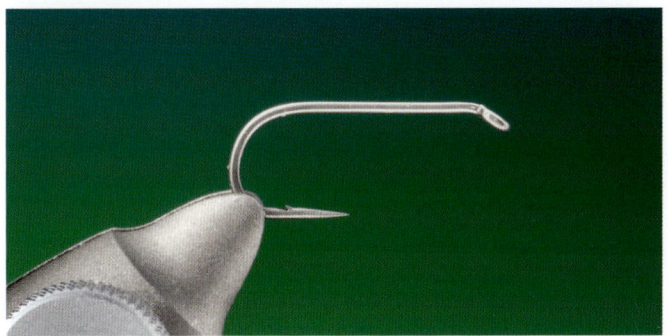

1 Secure your dry fly hook in the vice. Make sure that the hook shaft is horizontal. If you have a true rotary vice, centre the hook.

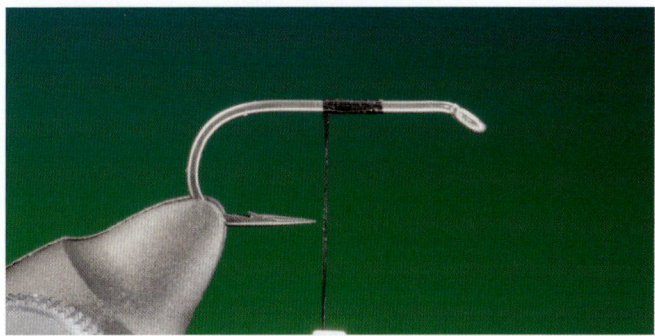

2 Attach your tying thread as shown and run a short foundation central on the hook shank. It's important that the front of the hook shank remains bare!

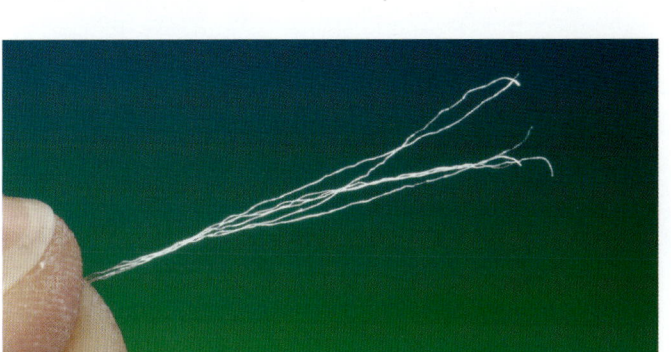

3 Take five or six strands of Antron sparkle yarn, no more.

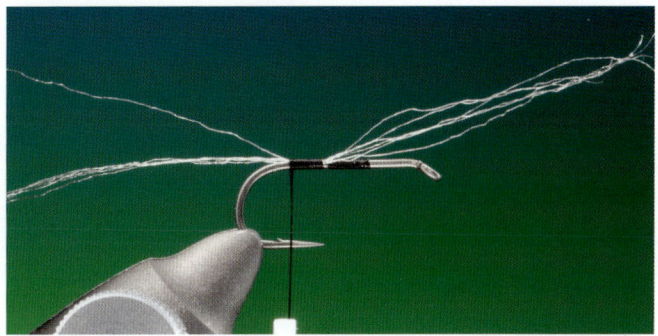

4 Tie this in as shown so they act as a tail at the rear of the hook.

65

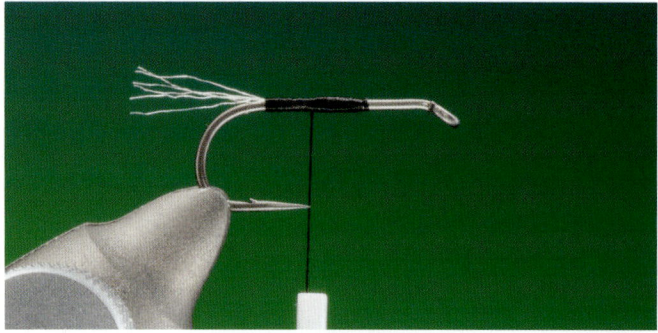

5 Trim off the surplus Antron yarn and tidy up with a few wraps of tying thread.

6 You will now need a larger length of Antron sparkle yarn. Tie this in on top of the hook shank.

7 As you secure one side of the yarn, pull it down, so it rests on the side of the hook shank as shown.

8 Repeat with the remaining length of yarn, on the opposite side of the hook shank. Secure with a few tight wraps of tying thread.

9 Spin a little dubbing onto your tying thread and build up a loose, buggy, swollen underbody. Again stop before you come to the front bare hook shank.

10 You now need to make three or four loose open spiral turns of tying thread into the hook eye.

11 Take hold of the Antron yarn at the rear of the hook and fold over forward. Secure with two loose turns of tying thread, tight into the hook eye. Distribute the Antron yarn around the hook shank.

12 Take hold of the two yarn ends and carefully draw them back towards the rear of the hook. If done correctly you will see the pupal shuck magically inflate!

13 Trim away the excess Antron yarn and secure with a few wraps of tying thread.

14 Cut and clean a small bunch of fine deer hair. Even up the tips in a hair stacker.

15 Tie in the deer hair wing over the body, approximately the same length as the hook shank. Make a few wraps of thread forward through the butts of the hair to secure it.

16 Trim away the surplus deer hair, taking care not to cut the wing material.

17 Apply a little loose dubbing to your tying thread, and dub a small scruffy head.

18 Once the head is done, make a couple of whip-finishes behind the hook eye.

19 Remove your tying thread and apply a tiny drop of varnish to the thread wraps.

20 The view of the emergent sparkle pupa from above, showing the wing shape and trailing shuck.

7

Hare's Ear Soft Hackle

Wood duck tail • Hare's ear body • Wet fly soft hackle

There is something special about soft hackled flies. For me, their very appearance screams: exquisite but deadly.

As the generic name suggests, the defining element here is the hackle. These of course, should be soft, and preferably a little downy. These quintessential hackles can be found in various forms and sizes, on most game birds. The most common hackles used are from the partridge, grouse and domestic hens.

Game birds and hens have what they call hard feathers. These have a shorter, tougher shaft. But the feathers themselves have a softer, tighter web woven within the barbs and barbules. You can easily test this. When you wrap a rooster hackle, the barbs will separate immediately as you wind it, standing 90 degrees from the hook shank without any grooming.

Conversely, the barbs on a hen hackle, when wrapped, are held together by the denser concentration of the softer barbules. These require persuading with a dubbing brush or comb in order to free them from each other. It is these characteristics that give the soft hackle fly its desired effect when fished.

This style of hackle will also absorb water quicker, helping the fly sink. When wet, the barbs will collapse over the body, trapping air. This style of fly doesn't have to look like a natural insect or other food, but rather it just has to swim and behave like something alive.

The amount of hackle used has always been a subject of heated discussion amongst flytyers/flyfishers. But heavily or lightly hackled, both have their advantages and you can decide what works for you.

The bodies of soft hackle patterns can vary enormously, from a smooth flashy tinsel, to an ultra buggy body as shown here, and everything in between.

Flashy tinsel and bright coloured floss bodies of various description, on soft hackles, are intended to add an attractor element to the pattern. Buggy bodies have a more organic appearance, resembling something more natural. You can also go for a subtle fusion of the two, as here: a buggy body and a flashy tinsel rib.

If you make the effort to to learn to tie just a handful of soft hackle styles, you can master them all.

This pattern can also be fished to great effect as a March Brown.

TECHNIQUES MASTERED

Wood duck tail
- Using wood duck flank barbs for a long swaying wet fly tail that swings when fished.

Hare's ear body
- Mixing hare's ear and mask hair and fur to make an extremely buggy dubbing body.

Wet fly soft hackle
- The correct method for selecting and wrapping a classic soft hen hackle for the best wet fly action.

Tying the Hare's Ear Soft Hackle

THE DRESSING

Hook: Mustad Heritage S80AP # 10-14
Tying thread: Sheer 14/0 brown
Weight: Optional
Tail: Wood duck
Rib: Oval gold tinsel
Body: Hare's ear dubbing
Hackle: Soft natural brown hen hackle

WATCH THE VIDEO

youtube.com/watch?v=dzjZdo0NIZk&t=14s

Tying the Hare's Ear Soft Hackle with Barry Ord Clarke

1. Secure your heavy wet fly hook in the vice. Make sure that the hook shank is horizontal. If you have a true rotary vice, centre the hook.

2. Attach your tying thread as shown, and cover the central area of the hook shank with a foundation.

3. You will now need a nicely marked wood duck flank feather. If you don't have wood duck you can use Mandarin duck flank or Mallard drake flank.

4. Cut a small long bunch of barbs from the flank feather, taking care that the tips are even. Try not to handle the bunch too much as this could put the barbs out of check, so they flare in all directions.

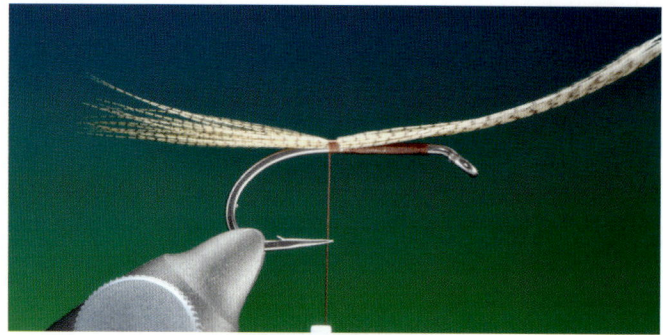

5 Place the bunch on top of the hook shank and measure the tail. Secure with a couple of wraps of tying thread.

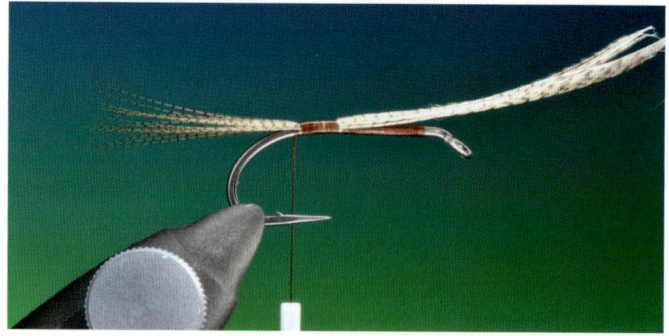

6 If you are happy with the tail position and length, you can tie it down, stopping just before you come to the hook bend.

7 You can now wrap forward securing the wood duck flank. Stop 3-4 mm behind the hook eye. Take care that you have enough room left for the hackle.

8 Use bright oval gold tinsel. Not all tinsel is the same so look for this type when buying: it makes a difference.

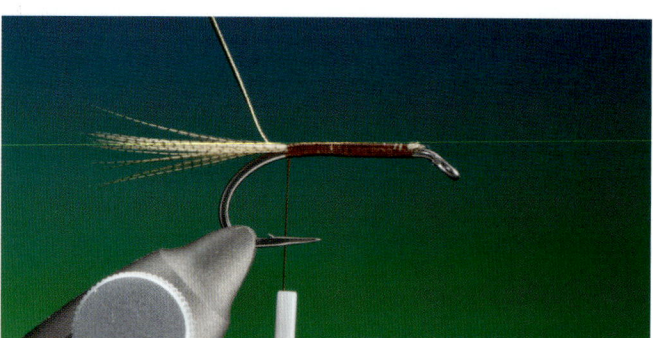

9 Cut a length of oval gold tinsel and tie this in the whole length of the hook shank, which results in a more even overbody. Finish at the rear of the hook shank at the tail base.

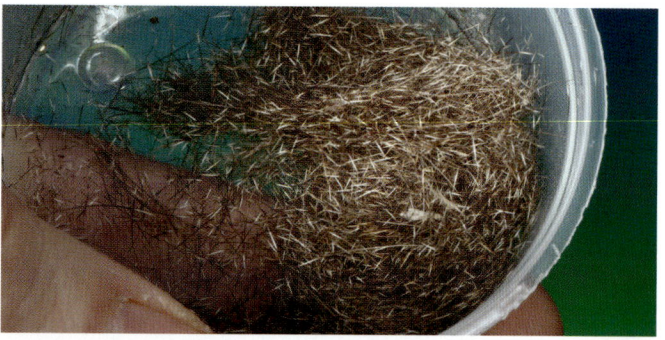

10 You will now need a nice blend of spiky hare's ear dubbing. See video for hare's ear dubbing; QR codes can be found in the Ubiquitous chapter.

11 Split your tying thread and make a dubbing loop at the very rear of the hook shank. Load the loop with the dubbing. Once loaded you can keep the loop tight and remove and adjust the dubbing before you spin it.

12 Spin your bobbin and form into a dubbing brush. Once tightly spun, you can use a brush to even out the fibres before you wrap it. Wrap forward in tight touching turns.

13 Take hold of your oval gold tinsel, taking care that it's not twisted, and make five or six turns of rib. Tie off.

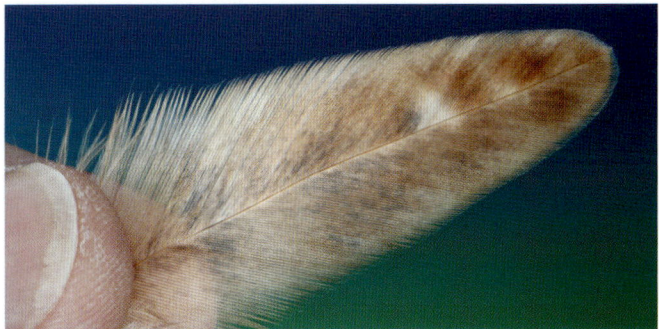

14 Select a nice soft brown hen hackle.

15 Remove the surplus tinsel. Tie in the hen hackle as shown, securing the stem.

16 Once you have tied down the stem forward, fold it back and secure once more.

17 Attach a hackle plier to you hackle and wrap the hackle forward folding each wrap backwards before the next. Tie off behind the hook eye.

18 Trim away the surplus hen hackle and secure with a couple of wraps of tying thread.

19 Flatten your tying thread by spinning your bobbin anti-clockwise and build a small neat head.

20 Make a couple of whip-finishes and remove your tying thread.

21 Finish by giving the head a few coats of varnish. Try not to get any varnish on the hackle.

8

All Fur Wet Fly

Lead wire flattening technique • Tapered fur body • Fur hackle

On a river, far away, a long, long time ago, before dry flies, nymphs and synthetics, wet flies were the only flies!

Today they have fallen somewhat out of fashion, and are kept alive by traditionalists who like to swing a wet down and across stream; and by those amongst us who know just how effective they are!

This simple technique for an all-fur wet fly is a modern twist on those early patterns. And as a fishing fly, if tied in several sizes, colours and weights, you don't really need many other sub-surface patterns. If any!

The addition of tapered lead wire adds not only weight, but sets the initial body shape for the pattern. This technique can be used on any pattern where such weight and body shape is desired. Achieving a fine, even taper on the lead wire is easy, if you know how. I suggest you see the video for the technique in full, as it's a little difficult to explain in text.

The trick to tying these all-fur soft hackles is primarily in choosing the correct fur and hair.

For the body, you need fur that is soft yet buggy, preferably with any long guard hairs removed. As with the rabbit I use here, it absorbs water well and has a subtle but lively action when wet. It's also available in more colours than just about any other fur product, whether natural or dyed.

I'm sure that all of you reading this have some zonker strips of rabbit or similar. If so, you have all that is needed. If you have a wire dubbing brush or dubbing rake, pull it along the length of a zonker strip a couple of times and I guarantee that you will have enough dubbing for one of these patterns. This also gives your zonker strips more movement, when the underfur is removed. It's a win-win situation.

Please note that it matters what denier tying thread you are using and what kind of fur you intend to spin in the dubbing loop. If you are using a light fine fur for the body, like muskrat or squirrel, splitting your tying thread for the dubbing loop will be sufficient. But if you employ a heavier more dense fur like rabbit, a split thread dubbing loop will more than likely break under spinning, so you will need to double your tying thread for these heavier, denser furs.

For the hackle, you will need a longer softer fur or hair that will have a flex and flow action when fished. If you have a soft hair that is too long, once again, as with the rabbit, all you have to do is load the Magic clip with the correct length of hair, before you cut off the remaining fur and tanned hide strip.

As shown in the last step image, you can tie this pattern with a rib and bead head for a heavier fly that will fish deeper, faster.

TECHNIQUES MASTERED

Lead wire flattening technique
- How to flatten and taper a short length of lead wire which when wrapped, shapes a tapered underbody.

Tapered fur body
- Using any natural underfur in a dubbing loop to create a tight dubbed buggy body for wet flies and nymphs.

Fur hackle
- Spinning fur or hair into a dubbing loop to make a very mobile fur wet fly hackle.

Tying the All Fur Wet Fly

THE DRESSING

Hook: Mustad Heritage S80AP # 10

Tying thread: Sheer 14/0 brown

Weight: Flattened lead wire, optional

Body: Bright green rabbit fur

Fur hackle: Natural squirrel body hair

WATCH THE VIDEO

youtube.com/watch?v=YhvmWy2ahAE

 Tying the All Fur Wet Fly with Barry Ord Clarke

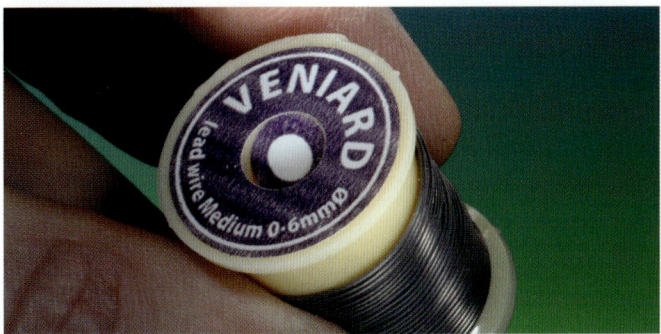

1 For weighting the fly you will need some medium lead wire.

2 Cut a short length, approximately 10cm long.

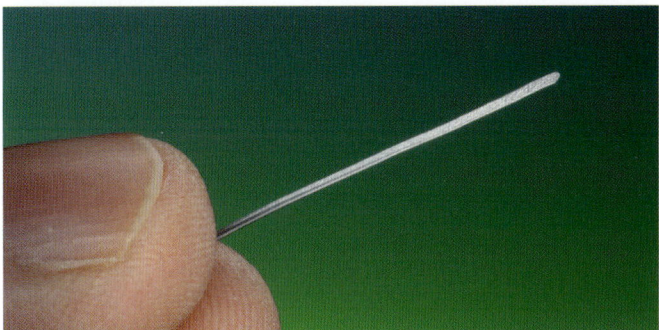

3 Watch the video to see the simple technique to flatten evenly one end of the wire into a taper.

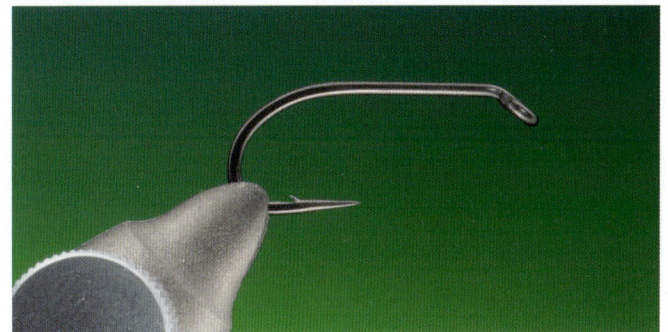

4 Secure your heavy wet fly hook in the vice. Make sure that the hook shank is horizontal. If you have a true rotary vice, centre the hook.

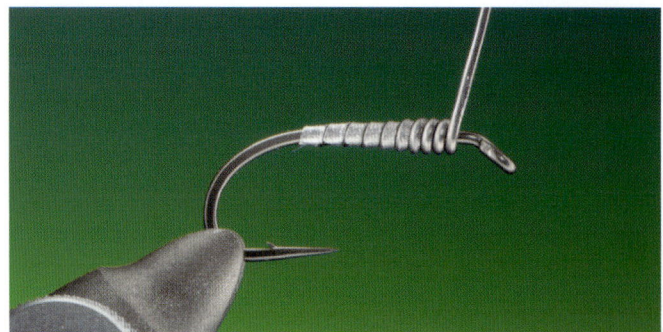

5 Wrap your tapered lead wire, starting with the flattened end at the rear of the hook and ending with the round wire a little behind the hook eye.

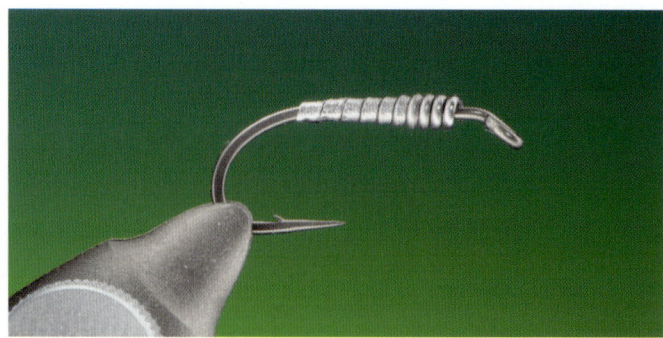

6 Break off the surplus lead wire and tighten the wraps by twisting it.

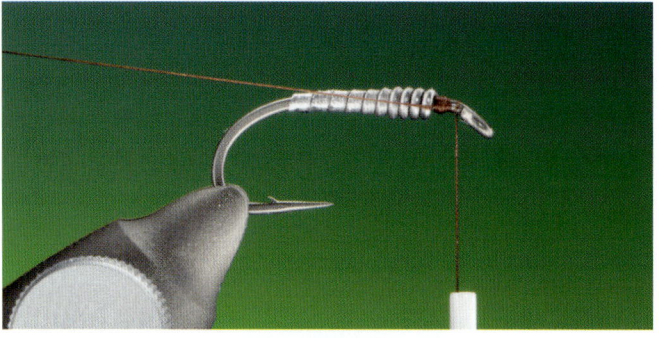

7 Attach your tying thread and form a stopper in front of the lead wire. This will stop it moving forward later.

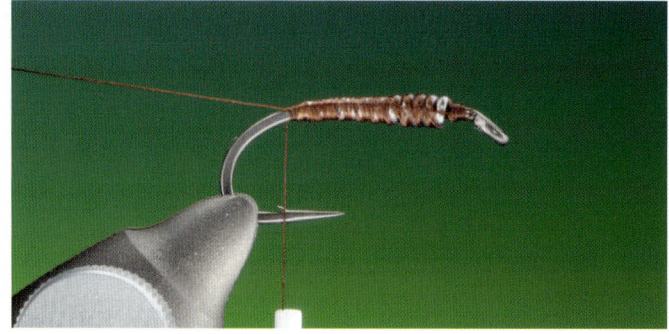

8 Cover the lead wire with tying thread and make a tying thread stopper at the rear of the lead wire.

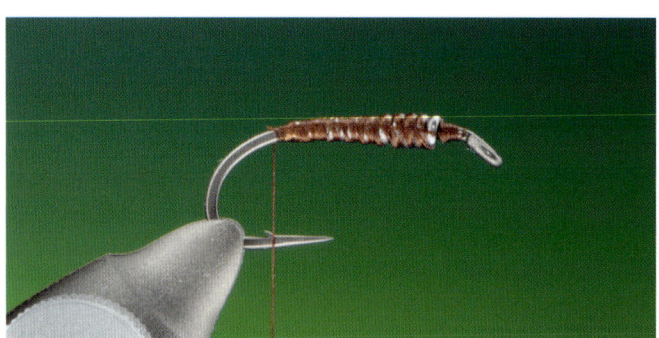

9 Remove the surplus tying thread.

10 Select some natural fur in the colour of your choice. Here I am using dyed rabbit fur. It's available in just about every colour and is perfect for this technique.

11 Double your tying thread and make a dubbing loop at the very rear of the hook shank. Load the loop with the fur. You can use a Magic clip for this if wished.

12 Put a heavy dubbing spinner on the loop and spin the fur into a dubbing brush. Once tightly spun, you can use a brush to release the fibres before you wrap it.

13 Catch the dubbing brush in at the rear of the hook and then wrap forward in tight touching turns over the lead wire.

14 Once you have covered the whole body of the fly, tie off the dubbing loop a little way back from the hook eye, and remove the excess. Give the body a brush to pull out any trapped fibres.

15 I have found that if you line the jaws of a Magic clip with thin strips of foam, it will hold hair, especially deer hair, much better.

16 If you have hair on a zonker strip, it makes the job easier when loading the Magic clip. Try and keep all the hair tips level; this will give a more even hackle.

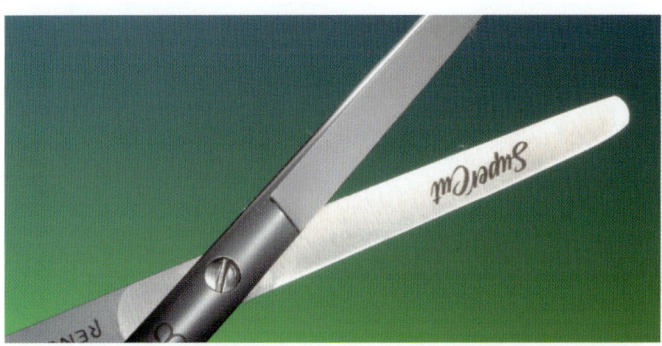

17 When cutting the fur strip, use long straight scissors. This will help keep the cut edge of the fur straight. If you use short scissors and have to make several cuts, you will get a staggered edge.

18 The loaded Magic clip with a straight cut edge.

19 Wax your tying thread, make a smaller dubbing loop and load the loop with your squirrel hair. Place a heavy dubbing spinner on the loop.

20 Now spin the dubbing spinner. Take care that you have spun it enough so the squirrel hair forms a hackle.

83

21 Once spun, give the hair a brush. Now when you wrap the fur hackle, comb back the fibres with each turn, as you would a standard wet fly hackle. Tie off behind the hook eye.

22 Build a neat head with tying thread.

23 Make a whip-finish and remove your tying thread.

24 Finish up by giving the head a few coats of varnish. Try not to get any varnish on the fur hackle.

25 View from below.

26 A heavier alternative, with bead head and wire rib, for when getting down faster is required.

Hare's Ear Parachute

Hare's Ear dubbing technique • CDC para post • Parachute hackle
---

Like its subsurface nymphal relative, no fly box should be without the Hare's Ear Parachute, in sizes 12, 14 and 16. This pattern can not only represent a whole load of mayflies duns, but its low scruffy profile just looks appetisingly buggy.

The hair/fur from the European brown hare's ears and mask is specified in many patterns. The mask and ears range in colour from pale tan, through muddy brown to almost black. The hair texture and length varies from fine soft underfur to almost bristle-like long spiky guard hairs. The ears are covered with a very close fine spiky hair with no underfur. A mixture of the fur and hair from the mask and ears gives the classic dubbing for the Gold Ribbed Hare's Ear and the March Brown.

Hare's ear dubbing can be blended and mixed with other synthetics, such as Antron and Ice dubbing. But, I prefer the unique qualities of 100% natural hare's ear dubbing, for much of the tying that requires it. The shorter and spikier, the better it is.

But the shorter and spikier it is, the more challenging it is to dub. The short stiff hairs won't stick to the tying thread. Using a little tying wax helps. But the best way is to split your tying thread and run a little wax over the loop. Then place your dubbing in the loop and spin it. This makes a very prickly, buggy-looking dubbing brush.

If you add a little underfur and/or mask hair, the dubbing is instantly made easy to use.

If you have a range of hare's masks, you can mix your own favourite shades of hare's ear dubbing. Everything from light to dark with varying grades of flash and bugginess.

I use an electric beard trimmer to remove the hair from the ears and mask, and a small electric coffee grinder to mix and chop it. Then I store it in small plastic soy sauce pots that you get with a sushi takeaway.

If you tie lots of flies, it's not only more cost effective to make your own dubbing, but you gain a greater understanding of the materials you are working with, and their applications. Something every serious craftsman should aspire to.

For the CDC parachute post I recommend natural untreated CDC feathers. These retain the natural oils and features that originally made CDC so popular through the founding flies of Charles Bickel and Louis Veya.

Sadly, much of the commercial product marketed as CDC today is far from the original. Washing and cleaning with strong detergents and colouring with hot dye baths render much commercial CDC void of any of the qualities that it may have had in it is natural state. Some suppliers even go so far as to impregnate treated and dyed CDC with a silicone to help it float again. So please buy wise when purchasing CDC, use your head and get your money's worth. It's better paying a little more for something that will do the job correctly, rather than something that is barely adequate!

TECHNIQUES MASTERED

Hare's ear dubbing technique
- How to make and best apply hare's ear dubbing for a spiky buggy result for both dry fly and nymphs.

CDC para post
- Selecting and using the correct amount of small CDC hackles to make a high floating, highly visible parachute post.

Parachute hackle
- An easy technique for preparing, mounting and wrapping a hackle that will result in the perfect parachute hackle every time.

Tying the Hare's Ear Parachute

THE DRESSING

Hook: Mustad Heritage R30 # 12
Tying thread: Sheer 14/0 tan
Tail: Dun or brown hackle fibres
Post: Natural CDC feathers
Rib: UTC Opal tinsel
Body/thorax: Hare's ear dubbing
Hackle: Chocolate dun

WATCH THE VIDEO

youtube.com/watch?v=GYA-dR3aM_o

Tying the Hare's Ear Parachute with Barry Ord Clarke

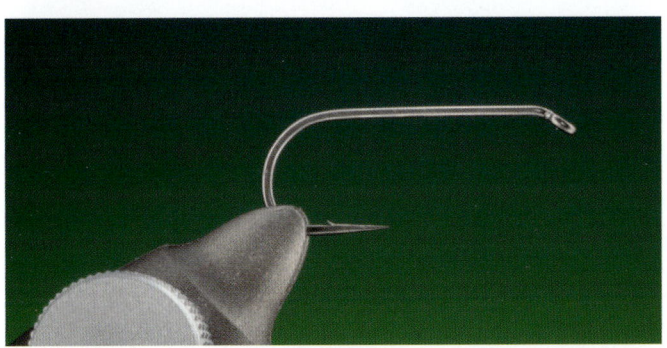

1 Secure your dry fly hook in the vice. Make sure that the hook shaft is horizontal. If you have a true rotary vice, centre the hook.

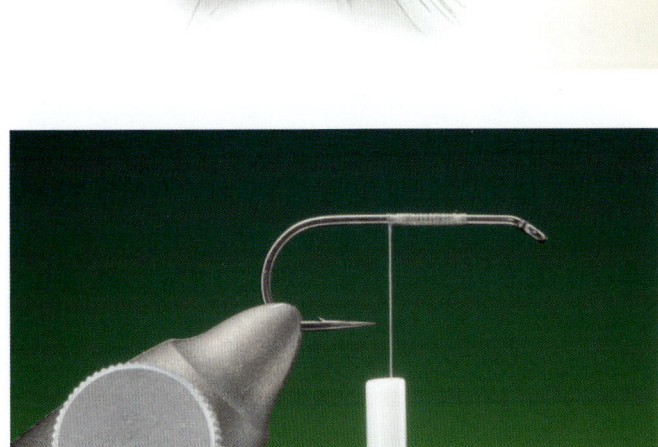

2 Attach your tying thread as shown and run a short foundation, central on the hook shank.

3 Cut a small bunch of dun or brown hackle fibres for the tail, then stack in a hair stacker to even the tips.

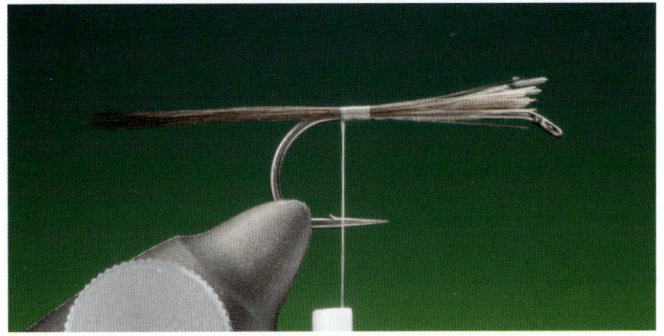

4 Measure the tail along the hook shank, and fix in the correct position with a few wraps of tying thread.

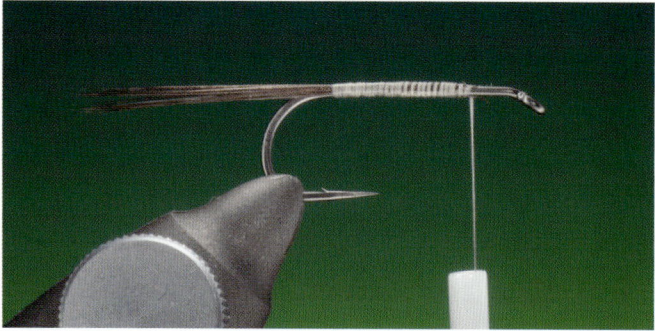

5 Flatten your tying thread by spinning your bobbin anti-clockwise, and secure the hackle fibres all the way up the hook shank.

6 Depending on the quality of your CDC feathers and the hook size being used, select the CDC for the post, normally one, two or three feathers.

7 Here I have three CDC feathers for the post. Level the tips of the feathers and measure the post before tying in.

8 Trim away the surplus CDC material at the rear, and cover with flat tying thread wraps.

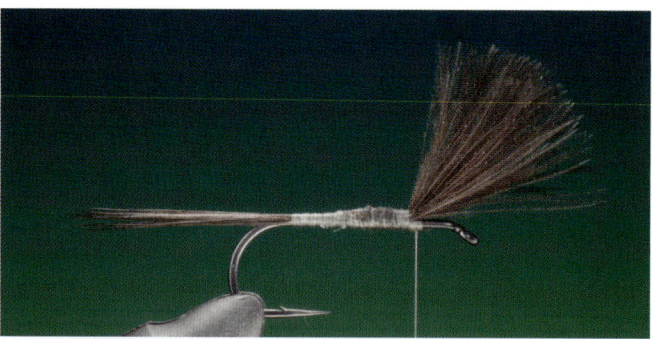

9 Once the post is secure, lift the CDC and make some supporting wraps tight in front of the wing to keep it vertical.

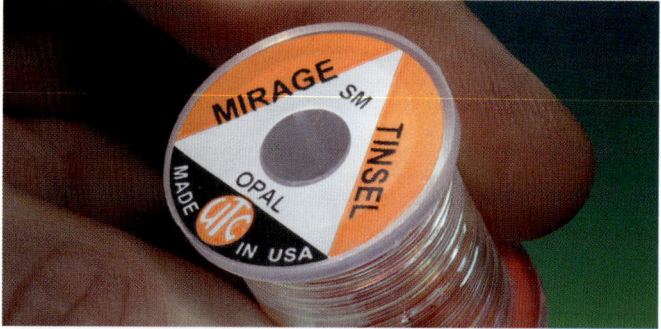

10 I like to use UTC, small opal tinsel for the rib. This has less weight that wire or oval tinsel, and lots of flash.

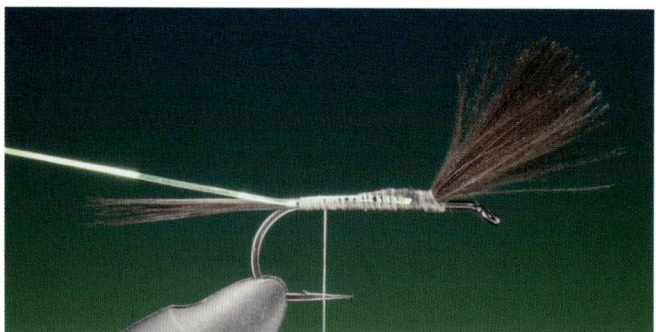

11 Cut a short length of tinsel and tie in the full length of the body. Finish at the tail base.

12 You will now need some fine spiky hare's ear dubbing.

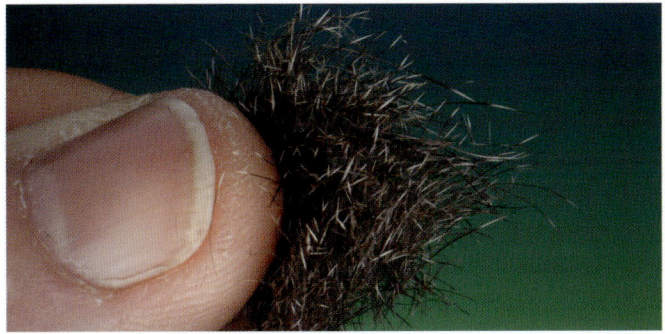

13 This dubbing when mixed should be super buggy and spiky. See my video on how to make it.

14 With your tying thread at the very rear of the hook, split your tying thread and make a dubbing loop. Wax the loop and load with hare's ear dubbing. Spin into a fine dubbing brush.

15 Dub the whole body, finishing just before the post base.

16 Grip the tinsel, and make four or five turns without twisting it and tie off at the post base.

17 Select and prepare a hackle. Here I'm using a medium chocolate dun, but brown works as well.

18 Support the post base with wraps of tying thread. Tie in the hackle at the post base, with a left – right, right – left cross as shown.

19 Secure the hackle stem up the post, and to the hook shank. Return your thread to the rear of the post.

20 Make another short dubbing brush with hare's ear dubbing and dub a thorax, a little thicker than the body.

21 Make a couple of whip-finishes behind the hook eye, and remove your tying thread.

22 Remove the hook from the vice and return vertical as shown.

23 Re-attach your tying thread securely to the post base.

24 Hang your tying thread/bobbin out of the way to the left on your vice. Attach a hackle plier to your hackle, and wrap in tight touching turns down the post base, into the thorax. Tie off.

25 Make a whip-finish or two around the hackle at the post base.

26 Remove your tying thread and return your hook to the horizontal position.

27 Bird's eye view of the finished Parachute Hare's Ear.

28 View from below.

10

Anorexic Mayfly

Coq de León comparadun wing • CDC extended body • CDC parachute wing

Before we get into the main tying instruction, I must explain the name I have given this pattern and the confusion surrounding it.

The original name that I gave this pattern is the Anorexic Mayfly, because of its sparse aspect. But as many of you have brought to my attention, the name given on my YouTube channel is different, the Emaciated Mayfly. Why? Bots scan auto-generated transcripts on the YouTube algorithm and flag individual words as harmful and hate speech, and anorexic is one of these. YouTube wouldn't allow me to publish the video with its original name. So you can call it whatever you like!

I designed this pattern in 2021 and although I haven't fished much with it to date, the times I have, it's produced way beyond expectations...

The simplicity of this design makes it possible to tie up mayfly duns of any size quickly with only a few basic materials and only one hook size. I recommend using large natural CDC feathers, but if these are not available, you can substitute the extended CDC body with either a cock or hen hackle used in exactly the same manner.

The front wing of true Coq de León may prove to be beyond budget or difficult to source. In this case, it can be substituted with Whiting Coq de León, (not to be confused with the real CDL) natural grey or pine squirrel tail or even fine Comparadun deer hair. If tied correctly, this pattern will splash down perfectly, float and behave just like the natural and provide a perfect footprint, time and again.

The key to tying the Anorexic Mayfly correctly lies in its proportions. One of the advantages of this pattern is down to its one hook size fits all. But this will only work if you accurately balance the amount of materials used. This is only learned and perfected by practice, but once mastered, the Anorexic Mayfly is quick and easy to tie.

The inspiration for this pattern was fuelled by Louis Meana Baeza's ingenious Pardon De Meana patterns and the best natural CDC available.

TECHNIQUES MASTERED

Coq de León Comparadun wing
- How to construct a Coq de León Comparadun-style wing for a high riding mayfly dun.

CDC extended body
- A simple semi-realistic and effective method for an extended mayfly body from a single CDC feather.

CDC parachute wing
- A half-wing, half-parachute technique for super delicate presentation using a split thread dubbing loop and natural CDC.

Tying the Anorexic Mayfly

THE DRESSING

Hook: Mustad Heritage C49S # 12

Tying thread: Sheer 14/0 tan

Wings: Coq de León and CDC

Extended body: A single large natural CDC feather

Post: White para post and wing

Thorax: CDC dubbing loop

WATCH THE VIDEO

youtube.com/watch?v=lsNBC12SEv0&t=57s

Tying the Emaciated Mayfly with Barry Ord Clarke

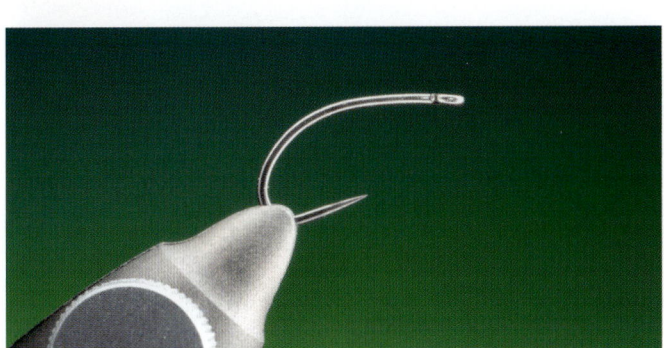

1 Secure your hook in the vice. Make sure that the hook shaft is horizontal. If you have a true rotary vice, centre the hook.

2 Select a fine CDL hackle with long barbs for the Comparadun wing. If you don't have CDL you can also use squirrel tail or fine deer hair.

3 Before you cut off the barbs from the CDL hackle, brush them back so they are 90 degrees from the hackle stem. You will need a bunch the diameter of a matchstick. Stick these in a hair stacker and even up the tips.

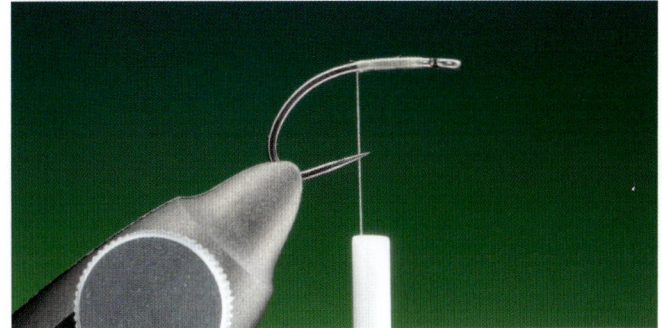

4 Attach your tying thread as shown and run a short foundation on the first half of the hook shank.

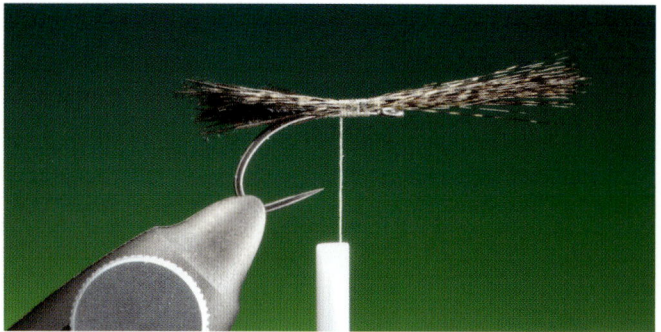

5 With the stacked tips facing forward, tie in the CDL bunch a little behind the hook eye, on top of the hook shank.

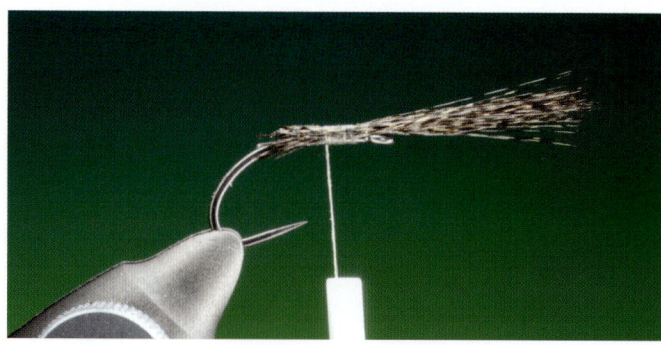

6 Trim the rear bunch of the CDL at a slight angle, so that they taper off towards the hook bend.

7 Now you must select a nice straight stemmed, natural CDC feather. If necessary, you can substitute the CDC feather with a cock or hen hackle.

8 Using straight scissors, trim the barbs of the CDC feather as shown. It helps if you use scissors with serrated blades. Take care that you retain two long barbs for the tail.

9 Again with fine sharp pointed scissors, cut out the centre hackle tip, leaving two tails. As far as I am aware, trout can't count, so two tails are as good as three!

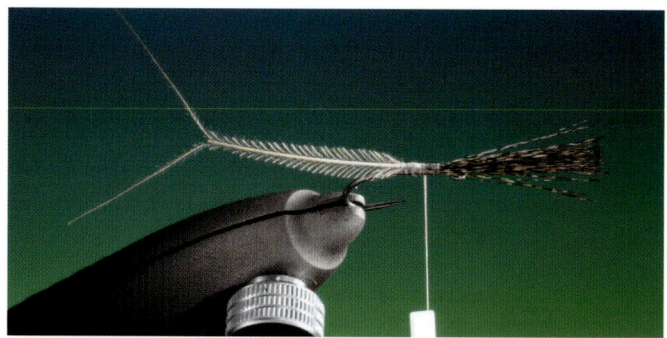

10 Measure the desired length of the extended body and trim, before tying in.

11 You will now need to cut a short length of para post, or similar material in the colour you favour. Take care that it's long enough to hold while you wrap the CDC parachute hackle.

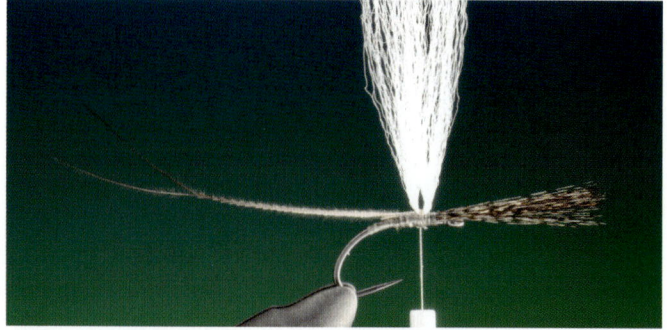

12 Tie in the para post material, centred between the CDL wing and the extended body.

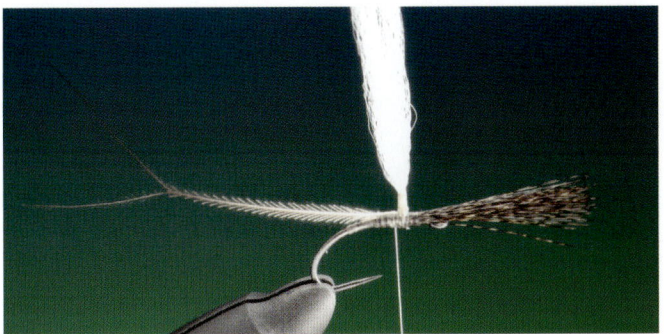

13 Make a few wraps of waxed tying thread around the base of the post to stiffen it. This will make wrapping the CDC parachute hackle easier.

14 Load a magic clip with a dense long-fibred natural CDC feather.

15 Split your tying thread into a dubbing loop and spin the CDC into a dubbing brush. Make two wraps around the hook shank between the post and the extended body.

16 Now wrap the CDC around the post base to make a parachute hackle.

17 Make another CDC dubbing brush. Take a couple more wraps around the post and then between the post and the CDL wing as a collar hackle.

18 Pull the CDL wing back and make the last few wraps of the CDC dubbing brush, tight into the CDL wing and then forward, finishing tight into the hook eye.

19 Your wing should now be vertical, 90 degrees to the hook shank.

20 Make a couple of whip-finishes and remove your tying thread.

21 You can now carefully trim the para post down to a slight sloping angle.

22 The para post should be trimmed like this. This functions as a fine quick sight indicator.

23 Rotate your vice and carefully trim away underside of the CDC so that it's a little more level.

24 It should be trimmed so that the underside of the fly will lie flat on the water's surface. This will give a lower riding presentation.

25 If desired, you can trim away the CDC fibres at the rear, each side of the extended body, leaving a wisp of long CDC fibres on each side.

26 The front view of the CDL Comparadun hackle, fanned out 180 degrees.

11

CDC Mayfly Dun

Moose body hair tail • CDC split wings • Tying thread rib

If you are lucky enough to fish hatches of the larger mayflies *danica* (green drake) and *vulgata*, you will know what a difference fishing a pattern with a good footprint makes.

The tail on this Mayfly Dun pattern has a significant part to play, not only for the footprint and silhouette as viewed by the trout, but it is also of great importance as the rear buoyancy aid and extension of the body.

Moose body hair is the perfect material for this. This hair is stiff and very coarse, with a large diameter at the base that tapers slowly to a superfine tip. It is extremely buoyant and between 2-5" in length.

These straight dark hairs make perfect tails. They float just as well as deer hair, but are stiffer and flare less when placed under pressure with the tying thread.

Setting the V-shaped vertical wings that are representative of so many upwinged patterns, is probably the most challenging technique here. The key to success is preparation. Take care that when selecting your CDC feathers for the wings, you use feathers that are of a similar size with square tips. *See step 7.*

Take heed when matching up the tips of these feathers for each wing that they are even and level.

You may have to pass them several times through your fingers until you get them lying right, but it's worth taking the time. *See step 8.*

Before attaching the wings, wax your tying thread which will hold them securely in place and stop them from slipping. Measure the wing along the hook shank. They should be approximately the same length or marginally longer than the shank. The wings should be attached, one at a time, to the hook shank, not on top but slightly to each side, at eleven and one o'clock.

Again, make sure that the left and right wing, when attached, are the same length. The excess CDC material at the rear of the wing should be trimmed at a slight angle, which will give you a foundation for the desired body taper.

Now you have to separate the wings. This you do by passing the thread around and between the divided wings, in such a manner as to catch all the CDC fibres together at the base of each wing.

The body of this pattern is not difficult to form, but occasionally flytyers have difficulty keeping it even and free of bumps and lumps. Here is the

key to this: before you start dubbing your body, twist only a little wisp of the dubbing nearest the hook shank. Once this top centimetre of dubbing is tight, make two turns of dubbing in between the tail base and rib material, to catch it in.

When the very first section of dubbing has been 'trapped' you can spin the remaining dubbing, but not too tight – this is important! *See step 18.* Continue dubbing the body, increasing the amount of dubbing slightly as you go, to build a taper.

For the rib I use double wine-coloured mono thread, to accentuate the body segmenting.

The only fishing strategy I need to mention for this pattern is drift it over a rising fish during a mayfly hatch. The fly will do the rest.

TECHNIQUES MASTERED

Moose body hair tail
- Using a little bunch of straight moose body hair for high floating extended mayfly tails.

CDC split wings
- Mastering CDC feather split vertical wings that create the perfect profile of an upwinged mayfly.

Tying thread rib
- How to rib the abdomen with a 3/0 coloured mono tying thread for accentuated segmenting.

Tying the CDC Mayfly Dun

THE DRESSING

Hook: Mustad Heritage R30AP # 12-10
Tying thread: Sheer 14/0 brown
Tail: A small bunch of straight moose body hair
Wings: Natural CDC feathers
Rib: UNI mono thread 3/0 wine
Body: Super Fine dubbing cream

WATCH THE VIDEO

youtube.com/watch?v=D6dlouJea7M

Tying the CDC Mayfly Dun with Barry Ord Clarke

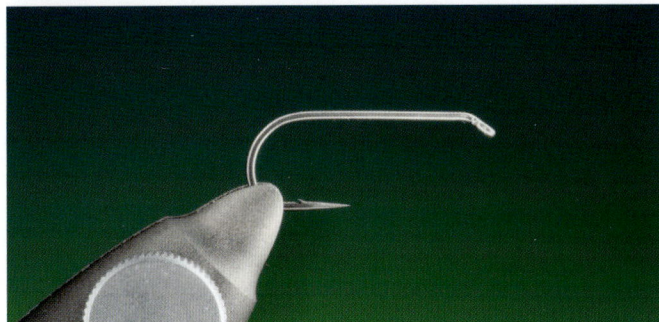

1 Secure your dry fly hook in the vice. Make sure that the hook shaft is horizontal. If you have a true rotary vice, centre the hook.

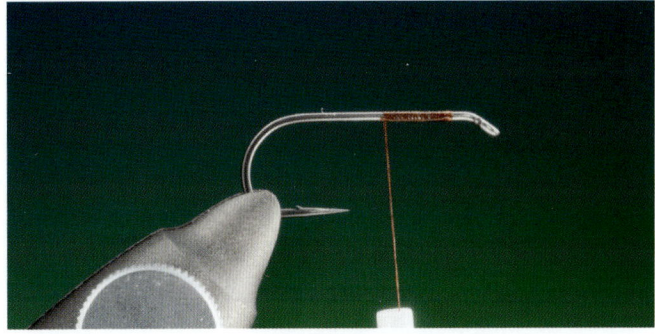

2 Attach your tying thread as shown and run a short foundation on the first half of the hook shank.

3 Cut a few dark moose body hairs, approximately 8-10. Clean them well by removing all the underfur and even the tips in a hair stacker.

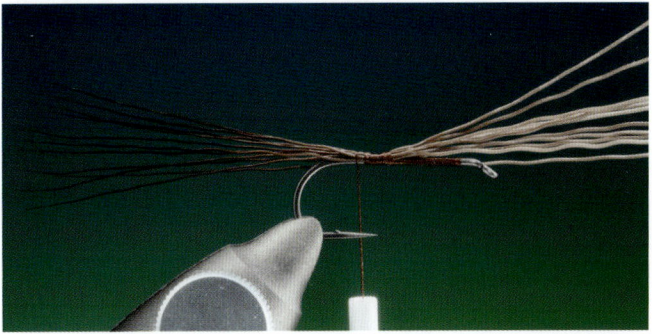

4 Take the bunch and measure the tail. For this pattern it should be double the hook shank length. Attach the bunch with a few hard wraps of tying thread. Flatten your tying thread and wrap loosely towards the tail base. This will stop it from flaring.

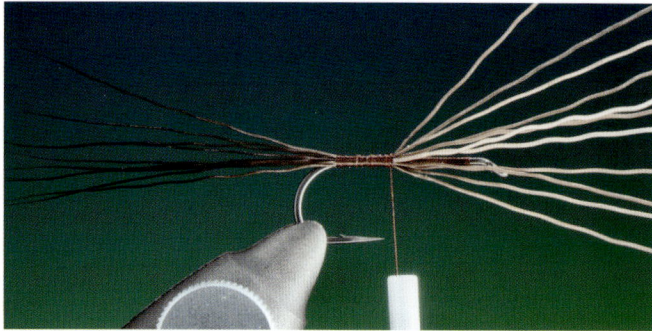

5 Keeping your tying thread with a flat profile, work your way back up towards the hook eye.

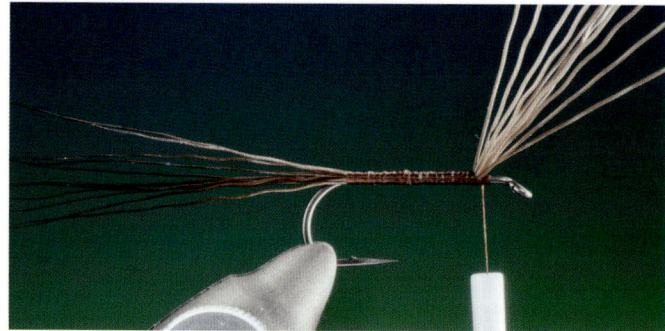

6 Once you have reached this position, make a couple of tying thread wraps in front of the hook eye. Trim away the surplus moose hair.

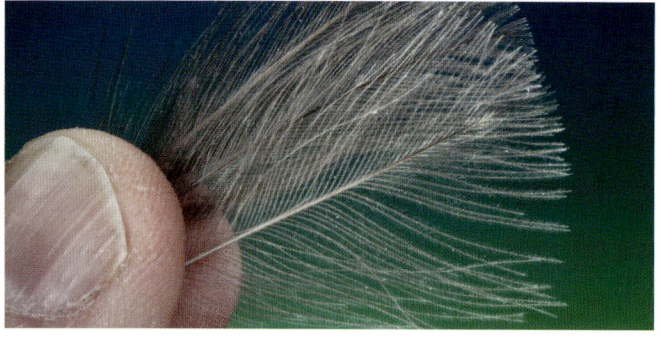

7 Depending on the quality of your CDC feathers and hook size being used, select the CDC for the wings.

8 Here I have two CDC feathers for the first wing. Level the tips of the two feathers before tying in.

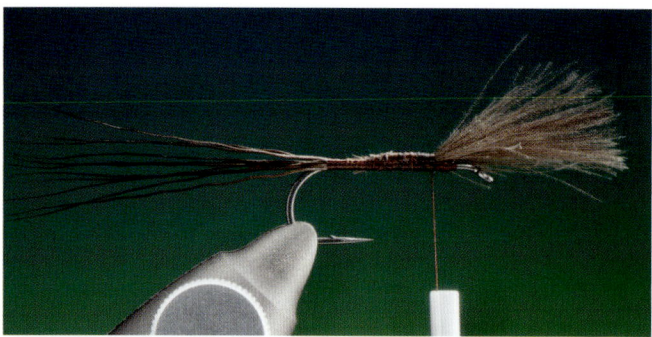

9 Measure the wing along the hook shank, and tie in slightly on the left side of the hook shank.

10 Once that the wing is secure, lift and make some supporting wraps in front of the wing to keep it vertical.

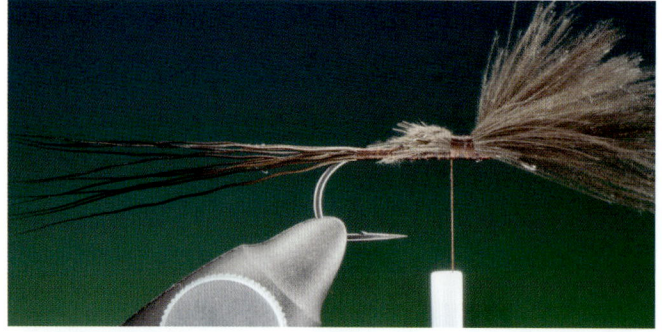

11 Repeat the whole process for the wing on the right-hand-side..

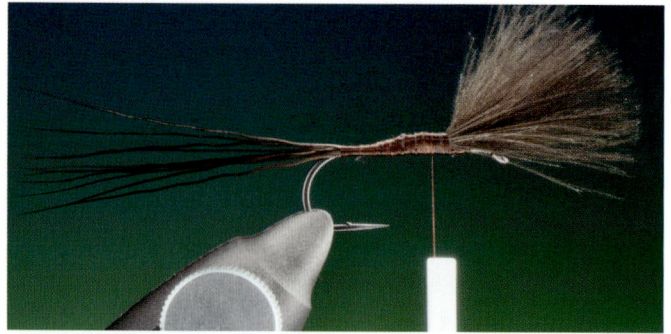

12 Using your tying thread, go over the body and make it taper.

13 With a figure-of-eight wrap, split the wings and support them with wraps of tying thread.

14 You will now need a short length of wine-coloured 3/0 mono thread for the rib.

15 Double the mono tying thread and tie in the whole length of the body, finishing at the tail base.

16 Now fold the doubled mono thread forward and make 2-3 wraps of tying thread over the tail base, so the thread is folded forwards.

17 Take a little Super Fine dry fly dubbing in cream for the body.

18 Attach a little dubbing to your tying thread and make two wraps between the tail base and the mono thread rib.

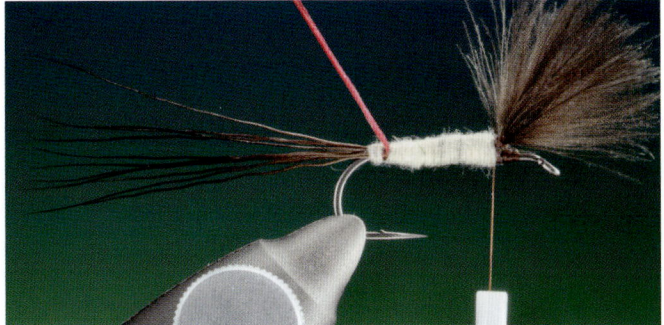

19 Spin more dubbing loosely on your tying thread, and start dubbing the body. As you dub the body, take care that you build a slightly increasing taper towards the wing. Finish at the wing base.

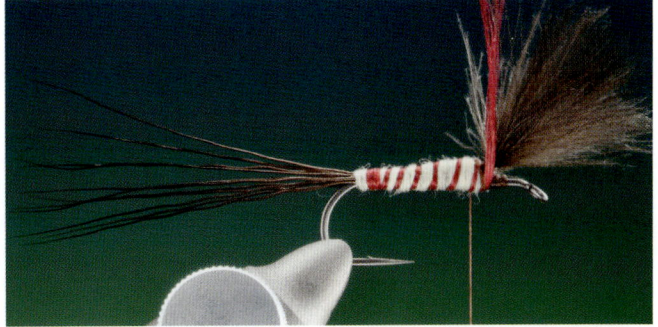

20 Take hold of your ribbing and just after the tail base make two wraps to form a thick band. Then wrap 5-6 times forward and tie off at the wings.

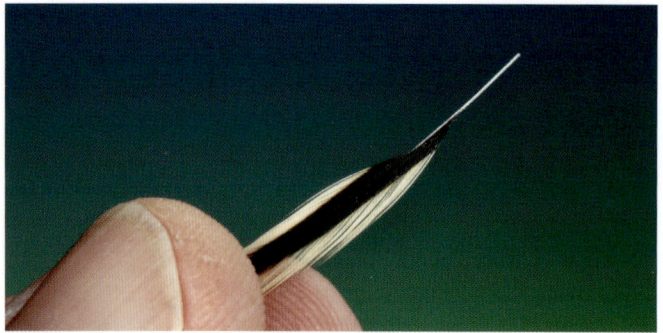

21 Select a golden badger saddle hackle for the collar. Prepare the hackle by stripping off the barbs as shown.

22 Tie in the hackle, where the body ends behind the wing.

23 Secure the hackle stem forward under the thorax.

24 Attach a hackle plier to your hackle and wrap forward, firstly behind the wing and then tight into the front wing and forward to the hook eye. Tie off.

25 Make a couple of whip-finishes and remove your tying thread.

26 The Mayfly Dun viewed from above with the hackle and split wings.

27 Front view of the hackle and wings. You can see that the hackle should be full.

28 Being an advocate of low profile presentation, I like to trim the underside of the barbs off.

Mallard Slip Wings

How to measure quill slip wings • Tying in slip wings
• Traditional style dry fly

Mallard slip quill wings have a long tradition in flytying history and, tied correctly, they can shape beautiful, well balanced, upright split wings, that create the perfect profile of an upwinged fly.

Here the flytying emphasis is not so much on the pattern, although it is a perfectly good one which, if colour and hook size are adjusted to match the hatch, will work in most mayfly situations. It's the *technique* shown for measuring and attaching the wings correctly which is the purpose of this tutorial.

Many flytyers, at all skill levels, can find tying in tidy matching slip wings for traditional dry flies somewhat challenging. As you can see from the online comments on the video, this technique has helped many flytyers master the problem. With a little practice the matching slip wings become relatively easy to tie.

Firstly, let's address the quills and which can be used. You can use the primary wing flight feathers from most waterfowl, although Mallard duck are

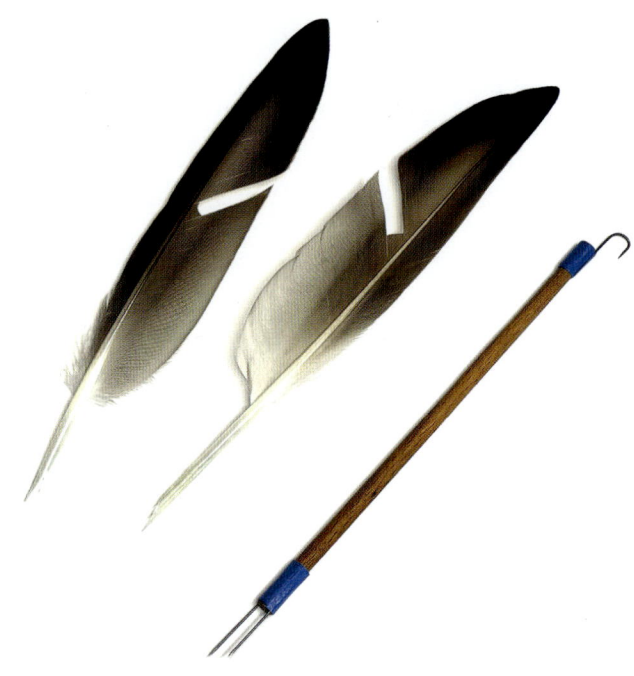

the default ones used in most quill wing patterns, unless others are specified in the recipe.

You will need two matching feathers, one from the right wing and one from the left. These are normally sold in set packs containing all the primary feathers from both wings.

If you are collecting these feathers yourself, the primary wing flight feathers are the first six or seven leading feathers from the tip of the wing and inwards, assuming the leading (longest) feather is number one. However, not all the quill on these primary feathers is suitable for quill wings. The lower quill section of the feather is generally too soft, and the upper quill section of the feather is too frail. Trying to use these sections will result in ruptured wing sections when you try to tie them in.

It's the middle third section of quill from each feather that is best suited for this technique.

There is a line on the rear of each feather that runs the whole length of the vane, which separates the soft section of the quill (barbs) from the hard section (tegmen). The hard section is nearest the shaft (rachis). *See step 9.* Your tying thread should at no time come down on this hard tegmen. This would cause the wing fibres to rupture!

Just as important as having matching right and left feathers, both quill segments should also match, and be taken from the same location on each feather.

Each quill section should also have the correct proportions, in respect to the hook size. This can easily be achieved with a small homemade tool.

I have made a number of these tools in the most common hook sizes that I use. Each of these has a hook in the same size that I will use, taped to the end of a small piece of dowel, *see step 3*.

As the width of the wing should be approximately the same as the hook gape, you can quickly gauge and section out a quill slip of the correct size with this, ready for cutting. Quill sections that are smaller than the hook gape are significantly easier to tie in, but they don't result in the desired wing profile.

These tools also work well for gauging wet fly wing segments.

Once you have paired the two matching quill sections, they have to be placed concave sides against each other and aligned at the tips and width. While holding both sections in your left hand, measure the wing length along the hook shank, gauging the length with the tip of your thumb. You can now secure the wings, with a upward pinch wrap. Here it's important to see the video!

TECHNIQUES MASTERED

How to measure quill slip wings
- How to make a simple measuring tool to make perfectly sized mallard quill wings.

Tying in slip wings
- An easy technique for preparing, pairing and attaching mallard quill slip wings.

Traditional style dry fly
- Wings, tail, body and hackle technique for tying a traditional style upwinged dry fly.

Tying the Mallard Slip Wings

THE DRESSING

Hook: Mustad Heritage R30AP # 12

Tying thread: Sheer 14/0 brown

Wings: Mallard wing quill sections

Tail: Coq de León

Body: 3 Moose mane hairs, choice of colour is yours

Hackle: Medium grey dun

WATCH THE VIDEO

youtube.com/watch?v=ioQKOVn2XQ0

Tying the Mallard Slip Wings with Barry Ord Clarke

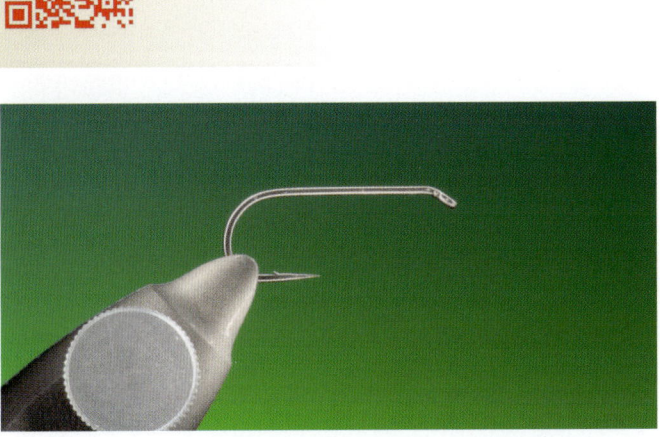

1. Secure your hook in the vice. Make sure that the hook shaft is horizontal. If you have a true rotary vice, centre the hook.

2. Select two matching right and left mallard wing feathers.

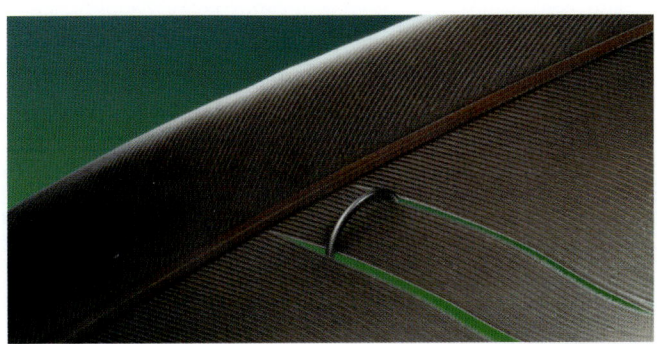

3. Now you will need a hook, the same size as you are tying on, to gauge the quill section.

4. This is done as shown, and the section should be as wide as the hook gape. Once marked, cut out the quill sections and put to one side.

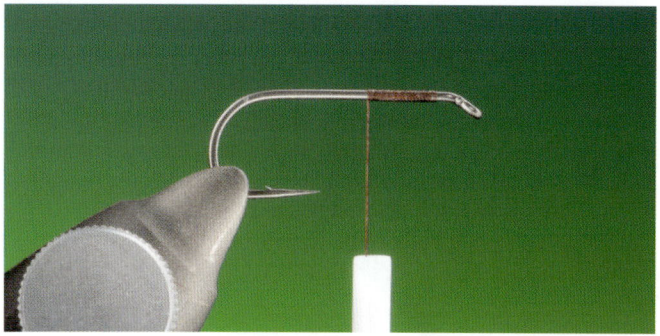

5 Attach your tying thread a little back from the hook eye and cover a small section of hook shank. Wax your tying thread.

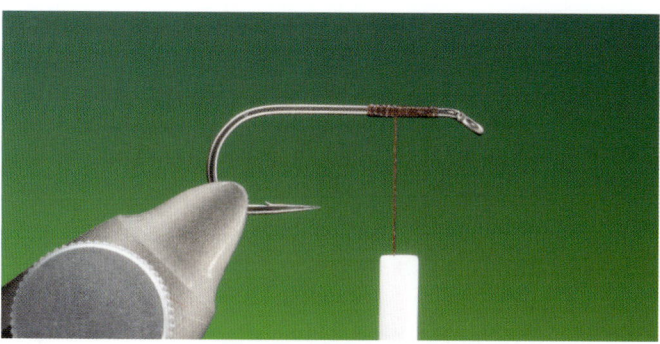

6 Run your tying thread back over the foundation, a little back from the correct wing position.

7 Take the two quill sections and make sure they are matching.

8 Now place the two quill sections, concave side against each other as shown. Take care that both the wing tips and width are aligned.

9 Here you can see the hard section (tegmen) of the quill. When tying in the wing, your tying threads should at no times come down on this. This would cause the wing fibres to split!

10 After you have measured the correct wing length with the hook shank, secure the wing with an upward pinch. *See Video!* This technique stops the wing from slipping and splitting.

11 Now you can release your hand and check the wings and position.

12 Secure the wings now with a few tight rearward wraps of tying thread.

13 Trim away the butt section of the quill wings at a slight angle. This will give the body the correct tapered profile later.

14 Spin your bobbin anti-clockwise, so your thread attains a flat profile. Now tie down the cut quill section towards the rear of the hook.

15 Keeping your tying thread flat, by spinning your bobbin anti-clockwise, build a nice tapered body. Then wrap forward, lift the wings and make a stopper tight into the wings to hold them in a vertical position.

16 You will now need a Coq de León hackle for the tail.

17 Cut a small bunch of Coq de León fibres and place them in a hair stacker, then even up the tips.

18 Measure the length of the tail and secure with a couple of wraps of tying thread. Check the length.

19 Secure the tail along the whole length of the body, still retaining the nice taper. Finish with your tying thread at the tail base.

20 Select three moose mane hairs. Here I am using one brown, one black and one grey. You are free to change the colour combination to match your hatch.

115

21 Align the hair tips. Once the tips are even, trim the hairs to the same length as the shortest hair. This is important for wrapping the hairs.

22 Tie in the three hairs by the tips. Secure into the tail base, taking care that all three hairs are tight into each other.

23 Secure the hairs, then run your tying thread forward behind the hook eye.

24 Take all three moose mane hairs, taking care that they are parallel and not twisted. Wrap the hairs together in tight touching turns, forward over the body. Tie off at the wing base.

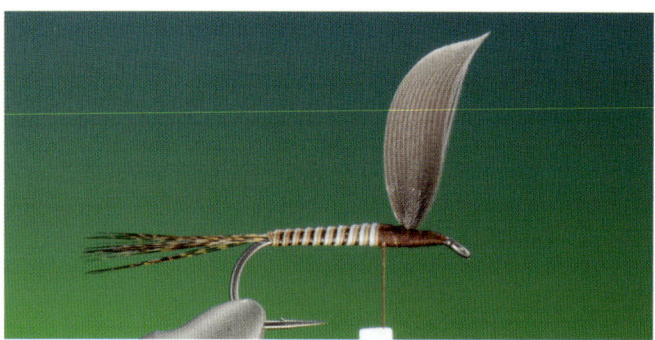

25 Trim off the remaining moose mane hair and secure with a couple of wraps of tying thread.

26 Prepare a medium grey dun hackle as shown, by stripping off the lower fibres. Take note that a little more fibre is removed from the wrapping edge of the hackle.

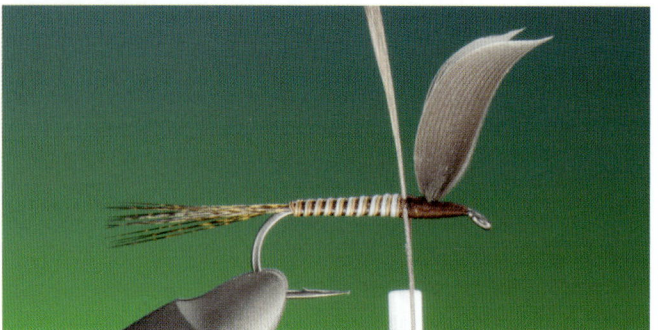

27 Tie in your hackle at the end of the moose mane body, 90 degrees from the hook shank.

28 Secure the hackle stem forward to the hook eye.

29 Attach your hackle plier to the hackle tip and wrap your hackle forward in tight even touching turns. Tie off a little behind the hook eye.

30 Whip-finish, remove your tying thread and, if desired, finish with a drop of varnish.

31 View of the wings from above, central and split.

32 Front view of the wing profile and hackle.

13

Comparadun

Deer hair Comparadun wing • Deer hair tail
• Classic dubbed mayfly body

The Comparadun is a pattern from Al Caucci and Bob Nastasi in their book, *Hatches*. This style of dry fly has a fan-like deer hair wing rather than a hackle.

The wing, which is situated a little back from the hook eye, should fan 180 degrees across the hook shank. This results in a very low riding presentation on the surface, something preferred by many anglers.

Having the right material for the wing is instrumental for success with this style of wing.

Some deer hair is packaged and sold as 'Comparadun' hair. This is normally from the hock and it comprises shortish, stiff hair with fine tips, that doesn't flare too much. But most flytyers will have some deer hair in their material collections that should work for this technique.

The hair originally recommended for this wing style is from the deer mask. The mask is the entire face area and ears of various species of deer (most commonly the white tail). As with a hare's mask, which we are all familiar with, the deer mask provides us with a huge range of natural shades from light tan to speckled brown to light dun and more. It also furnishes us with a huge amount of hair textures from short, stiff and ultra-fine, to coarse, all in varying lengths for flies from a size 4, down to the tiniest patterns. For the caddis fly specialist, a deer mask is an absolute treasure.

Deer masks were relatively easy to obtain when I started tying flies and were carried by many large suppliers, including Orvis. Today however, they are the proverbial unicorn amongst tying materials, and are hardly possible to find for sale.

Be that as it may, it is always possible to prepare your own deer mask! Removing the mask from a deer, if done correctly, is a precise and somewhat time-consuming task. If you are a hunter, or know a deer hunter, its well worth asking if they can obtain the head from a late season doe. Avoid bucks with even, small antlers, as they are more problematic to skin.

For the full skinning and preparation instructions video, scan the QR Code.

The key to mastering the Comparadun correctly, lies chiefly in setting the wing. As mentioned briefly earlier, you need deer hair that is a little finer and stiffer and that won't flare too much. Avoid using that sold as spinning hair. I prefer hair that has nice, fine, barred tips, which gives the finished fly not only a more aesthetic touch, but also highlights the desired wing profile.

It helps here if you wax your tying thread and the short foundation of thread on the hook shank. This will help your thread grip the deer hair and stop it from slipping around the hook shank.

Cut a bunch of hair, initially a little larger than you believe you require. Once you have combed out the underfur and shorter hairs, it will be reduced slightly. And it's always easier to remove hair than add it!

Once your hair bunch is cleaned, you can stack it. This again is important for the desired wing profile.

When the tips are evenly stacked, the hair bunch can be carefully removed from the stacker and placed on top of the hook shank to measure the wing length. It should be approximately the same as the hook shank.

You now move the bunch forward to the wing position, with the tips pointing out over the hook eye. With a few wraps of tying thread, secure the bunch, while still holding the hair bunch with your left hand. When the hair is firmly fixed, you can trim away the surplus as neatly as possible and then tie down the butt ends.

You can set the wing now, but here I tie in the tail and dub the body. I find that finishing dubbing the body, tight into the wing, almost pushing it forward, packs the deer hair and makes for a tighter wing when finished. You can see the alternative technique in the video.

Now, using your thumb nail, push your wing back 90 degrees from the hook shank and wrap your tying thread forward, through and in front of the wing. With a few wraps of tying thread, make a stopper in front of the wing to hold it up vertically.

Pull the wing up from the underside of the hook so it's in the correct fan shape. Apply a tiny amount of dubbing and cross over the underside of the wing *as in step 15*. Dub a little more, tight into the front of the wing, and whip-finish.

TECHNIQUES MASTERED

Deer hair Comparadun wing
- How to select and prepare the deer hair and mount a Comparadun wing.

Deer hair tail
- Tying in a deer hair tail for a classic dry fly, without it flaring.

Classic dubbed mayfly body
- How to apply and dub a dry fly body for a mayfly dun.

Tying the Comparadun

THE DRESSING

Hook: Mustad Heritage R30AP # 12
Tying thread: Sheer 14/0 grey
Wing: Fine natural deer hair
Tail: Fine natural deer hair or elk hair
Body: Super Fine dubbing Adams grey

WATCH THE VIDEO

youtube.com/watch?v=68iyPkK8VLY&t=481s

 Tying the Comparadun with Barry Ord Clarke

 Deer mask prep

rb.gy/jeldem

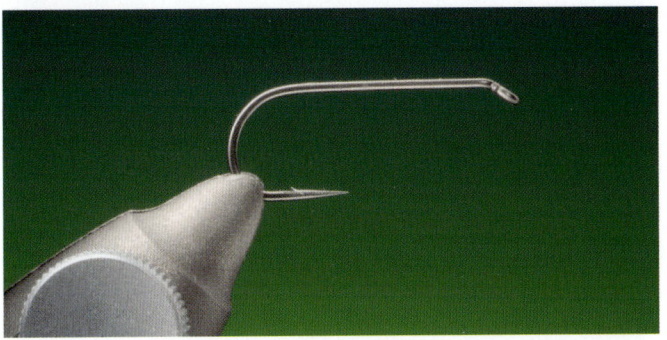

1 Secure your hook in the vice. Make sure that the hook shaft is horizontal. If you have a true rotary vice, centre the hook.

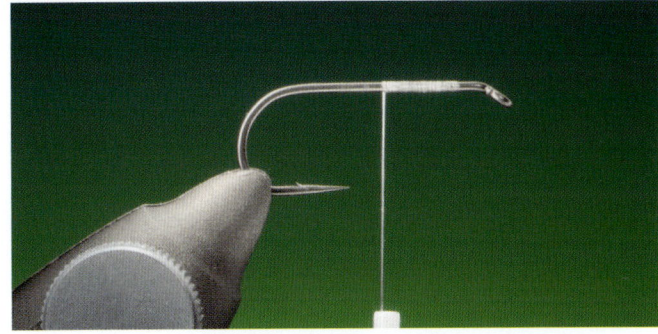

2 Attach your tying thread as shown and run a short foundation on the first third of the hook shank. Wax your tying thread and the foundation.

3 Cut a small bunch of deer hair and remove all the underfur and shorter hairs.

4 Once the hair is cleaned, place in a hair stacker and even up the tips. If you have used deer hair with barred tips they will now be stacked nicely.

5 Remove the deer hair from the stacker and measure the wing along the hook shank. You now move the bunch forward to the wing position and tie in with a few wraps of tying thread.

6 If you are happy with the wing, fix it firmly. Then trim away the surplus deer hair as shown, at a slight angle.

7 Tie down the butt ends and run your tying thread back towards the rear of the hook.

8 Cut a small bunch of fine deer hair or elk, clean and stack as before. You don't need much here, just a small bunch.

9 Measure the tail and secure with a couple of wraps of tying thread. If you are happy with the tail length, tie it down. Take care that when you approach the tail base, you slacken off your tying thread. If you apply too much tension here, the tail will flare.

10 Trim away the excess tail material and spin your bobbin anti-clockwise. This will give your thread a flat profile. Now build a slight taper over the body with tying thread.

11 You will now need some grey dry fly dubbing for the body.

12 Wrap your tying thread back to just before the tail base and apply a wisp of dubbing. Spin the dubbing at the very top of your tying thread only, and make a couple of wraps to catch it in.

13 Once your dubbing is caught in, you can spin the remaining dubbing tight and dub the whole body. Finish the dubbing tight into the wing, almost pushing it forward.

14 Using your thumb nail, push the wing back so it adopts a vertical position. Move your tying thread forward through the wing and make a stopper tight into the wing to hold it in position.

15 With a just a little more dubbing, make a cross dubbing under the wing to pull each side of the wing up, into the correct fan position.

16 Finish with a little more dubbing tight into the front of the wing. This will help it hold its vertical position. Whip-finish and remove your tying thread.

17 A view from above of the finished Comparadun wing position.

18 From the front, the deer hair Comparadun wing should fan out 180 degrees, so the lower left and right wing will support the fly on the water.

Clarke's Caddis

Elk hair extended body • Elk hair down-wing
• CDC dubbing brush wing and hackle

This somewhat unconventional deer hair technique is made easier if long hair can be used. Although regular long deer hair can be employed, I prefer to use the stiffer waxy cow elk hair. The extra length and rigidness of this hair makes the handling and technique much easier to perfect.

The North American elk (*Cervus canadensis*), is the second-largest species of deer in the world, also known as wapiti in Europe. Elk, on the other hand, in Europe is the name given to the European moose (*alces alces*). The two are very different and should not be confused!

Elk provides us with perhaps the broadest range of hairs to meet nearly every tying situation or requirement.

When buying cow elk hair, look for 'Cow Elk Rump' hair that is long and straight, with steeply tapering tips with fine markings. You can see that it looks and feels a little waxy to the touch.

These larger diameter, late season elk hairs are already extremely buoyant to start with. But when folded to form the extended caddis body, they create a little elk hair pellet that traps air. The length or size of the extended body can be modified to the hook size being used, if required.

When stacking the elk hair, use a larger hair stacker to keep the long hairs in check. You may find that the hair you have is slightly curved. If so, hold your hair stacker at a 45 degree angle when rapping it on the table. At once all the hairs fall into line with each other. If you have managed to get some cow elk with unbroken tips and vivid markings, after stacking, the tips will align to beautiful effect.

Finishing the collar of the fly with the CDC dubbing brush can be done sparsely or full, depending on how you wish to fish the pattern, either as a high riding dry or as an emerging pupa.

The hook size and style plays a significant role in this pattern. The size 14 C49X hook that I use on the one shown here, has a deep gape. This not only retains a good 'bite' between the hook point and extended body, but also functions as a keel, keeping the fly fishing in the correct way.

This I fish, well dressed with floatant, on a floating line in combination with a slow sink Poly Leader. This leader modifies your floating line into an extremely slow sink tip. This is still light enough to keep your fly dry, if required, but will make it dive and swim just under the surface when you make a retrieve. As your retrieve ends, the fly 'plops' up in the surface again, imitating an emerging caddis pupa.

TECHNIQUES MASTERED

Elk hair extended body
- A technique for making an extended body from a small bunch of cow elk hair.

Elk hair down-wing
- Elk hair down-wing made from the same bunch of cow elk hair as the extended body.

CDC dubbing brush wing and hackle
- A split tying thread dubbing loop loaded with CDC fibres to form a CDC dubbing brush for the hackle and wing.

Tying Clarke's Caddis

THE DRESSING

Hook: Mustad Heritage C49XSAP # 14
Tying thread: Sheer 14/0 tan
Body: Natural cow elk/deer hair. Late season
Shell back: Natural deer hair

WATCH THE VIDEO

youtube.com/watch?v=1OV3HV80wJ0&t=339s

Tying Clarke's Caddis with Barry Ord Clarke

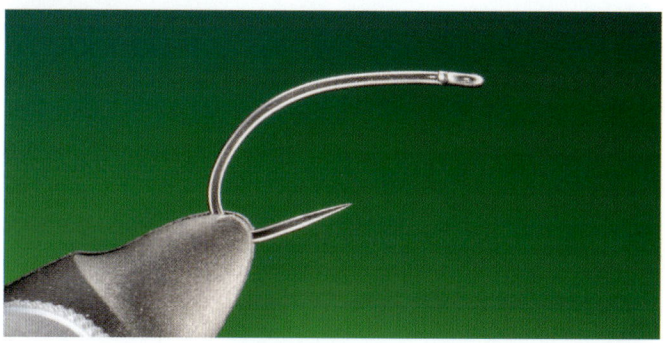

1 Secure your hook in the vice. Make sure that the hook shaft is horizontal. If you have a true rotary vice, centre the hook.

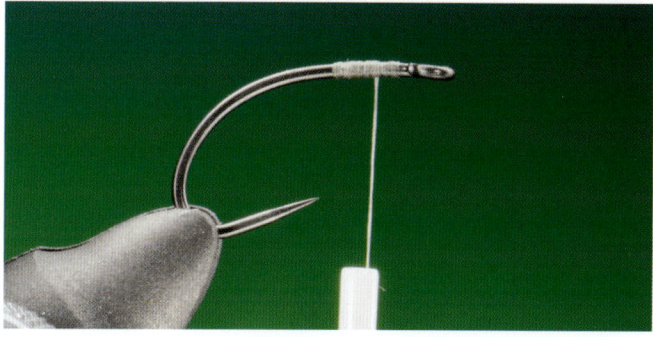

2 Attach your tying thread as shown, a little behind the hook eye and run a foundation a short way along the hook shank.

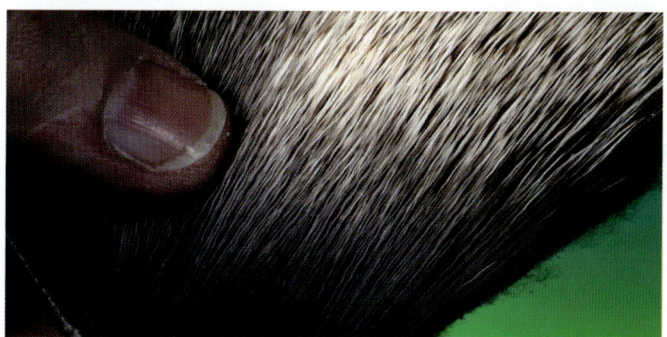

3 You will now need some fine, well-marked cow elk hair. It's not necessary, but it helps if it is long hair. This will make handling it easier. You can also use a standard deer hair.

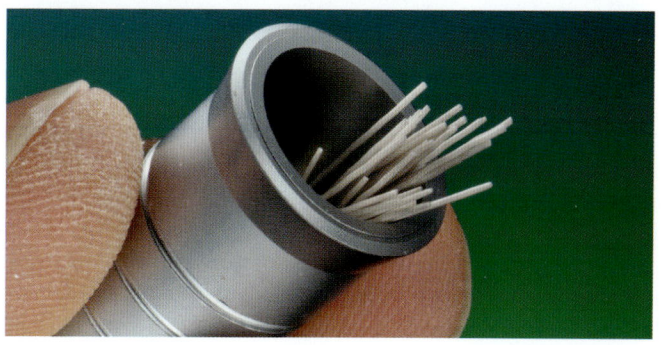

4 Cut a small bunch of the cow elk hair, keeping it as long as possible. Now stack the bunch in a hair stacker to even up the tips.

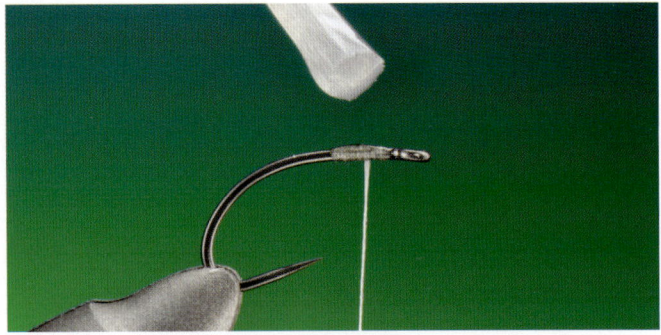

5 Before you tie in the elk hair, place a small drop of super glue gel on the short tying thread foundation and a little down the tying thread.

6 Once the elk hair is stacked and the tips even, measure the wing as shown over the hook shank.

7 Remove the hair from the stacker and carefully secure with a few wraps of tying thread while still holding the rear ends of the hair.

8 You will now need a monofilament loop. You can make these yourself from regular mono or use a dental floss threader as shown here.

9 Place the mono loop over the rear bunch of elk hair. Take care to keep the hair in the bunch tight, even and parallel.

10 While holding tension in the mono loop with your left hand, fold over the elk hair bunch and secure at the wing base with a few tight turns of tying thread. Pay attention to the length of the extended elk hair body.

11 When the extended elk hair body is secure at the wing base, you can remove the mono loop from the body.

12 Carefully wind your tying thread through the spun elk hair and wing, taking care not to misplace any of the wing fibres. This will fasten the body and wing securely. Finish with your tying thread behind the hook eye.

13 With super-fine sharp scissors, trim away the surplus elk hair as tight as possible between the wing and extended body.

14 Wax your tying thread before you fold back the elk hair wing tips and tie down as shown. The wax will help your tying thread grip the elk hair and stop it from slipping.

15 You will now need a nice long fibered natural CDC feather for the dubbing loop.

16 Load a small Petitjean Magic clip with the CDC fibres. Leave only a short amount protruding from the clip jaws.

17 Spin your bobbin anti-clockwise so that your tying thread gets a flat profile and split it to form a dubbing loop. Spin the CDC fibres in the loop to form a CDC dubbing brush. Make a couple of turns of CDC between the body and the wing.

18 Continue wrapping the CDC dubbing brush forward over the head of the fly and finish with your tying thread behind the hook eye.

19 Make a whip-finish or two and remove your tying thread.

20 Carefully trim away the underside of the CDC fibres as shown.

15

Diving Caddis

Neon hot spot • Mallard wing • Partridge body hackle

Caddis flies generally lay their eggs in late afternoon or evening and when doing this, they either flutter, crawl or dive. They flutter and bounce over the surface of the water, depositing their eggs each time that they splash down. The females of the crawling and diving species enter the water, as their name suggests, then lay their eggs on the subsurface structure. You can see the full-page photo of my caddis egg-covered waders, after a successful evening's fishing with the diving caddis.

I feel it's worth quoting here a passage from the 1955 book, *How to Fish from Top to Bottom* by Sid Gordon. He noted the importance of the egg-laying diving female caddis flies.

'As it pierced the water film, an almost unbelievable change came over that drab fly. It suddenly seemed to be encased in a bright gleaming bubble, so bright that it looked like a shining ball of quicksilver. Fast moving legs propelled the bubble, angling it down toward the bottom of the clear water.'

'There the bubble crawled nervously over a rubble stone and stopped. What was that fly doing there? I watched the bright, silvery insect for a moment or two, then reached down to seize it, but the fly let go its hold upon the rock and drifted away.'

Although the pattern illustrated in this chapter is unweighted, except for the weight of the heavy hook, you can tie it with weight under the dubbed body, or even for deeper fishing patterns, very effectively with a bright silver bead head.

I fish this pattern on its own, dressed with a little CDC oil floatant. This traps the air better, between the fibres of the body dubbing and veiling. I also use a short medium fast-sinking Poly leader. This pulls the fly 'down' on the retrieve, and lets it 'pop up' again, when you pause the retrieve.

Hen mallard breast feathers used for the wings can be a little difficult to source from tackle shops, but Veniard do carry them as standard. The mallard is the most common duck species, so if you know a wildfowler, or even a game butcher, you should be able to pick up a whole bird/skin, for next to nothing. For the caddis fly specialist, they are a must!

You can also use, as an excellent substitute, the chestnut brown mallard drake neck feathers, as used in the Welshman's Button on page 149.

Feathers from the Grey, Hungarian or English partridge, as it's also known, are essential if you are tying trout flies. The speckled brown back and mottled grey breast feathers are used in many traditional and modern patterns, both wet and dry. When buying partridge feathers, please try to avoid packs stuffed with feathers. These have a minimal amount of usable feathers, many of them twisted and warped out of shape from being mishandled and stuffed into a small plastic bag.

If you have the budget, buy a whole skin. On a cautionary note, if you are going to purchase a whole skin, please don't do it online. From my own experience, once ordered and paid for, anything can turn up in the post!

You have to handle and scrutinise any natural material that you are going to purchase, especially one of a substantial investment.

TECHNIQUES MASTERED

Neon hot spot
- An easy technique, using UNI Neon thread for making a hi-vis hot spot and ribs for the egg-laying female caddis.

Mallard wing
- Two hen mallard breast feathers put to good use as a caddis fly wing.

Partridge body hackle
- Using a speckled partridge feather as a body veiling and how to hold it in position with dubbing.

Tying the Diving Caddis

THE DRESSING

Hook: Mustad Heritage S80AP # 12-16

Tying thread: Sheer 14/0 brown

Rib/butt: UNI Neon orange 1/0 tying thread

Body: SLF dubbing

Wings: Two hen mallard breast feathers

Hackle: Speckled partridge feather

Head: Super Fine dubbing

WATCH THE VIDEO

youtube.com/watch?v=1yh1IKYAlXM

Tying the Diving Caddis with Barry Ord Clarke

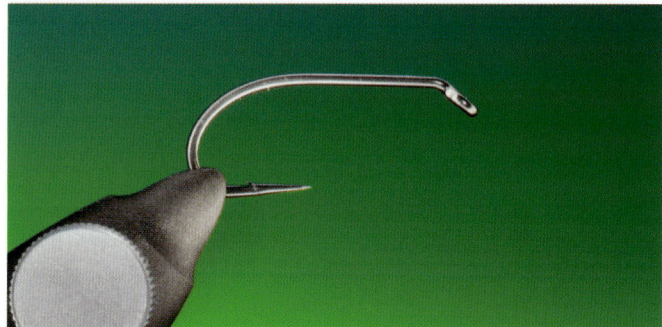

1 Secure your wet fly hook in the vice. Make sure that the hook shaft is horizontal. If you have a true rotary vice, centre the hook.

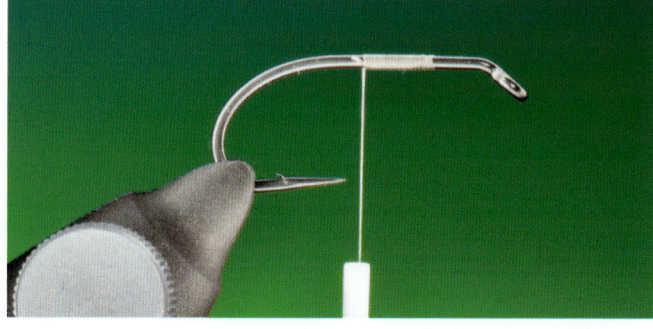

2 Attach your tying thread central to the hook shank and run a short foundation.

3 You will need a short length of UNI fluorescent orange neon tying thread for the hot spot and rib.

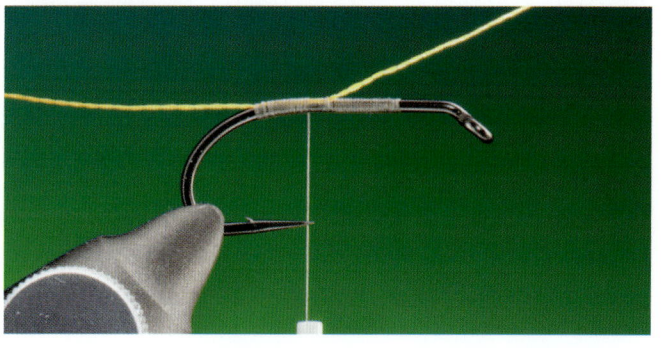

4 Attach your neon tying thread as shown, a little down the bend of the hook.

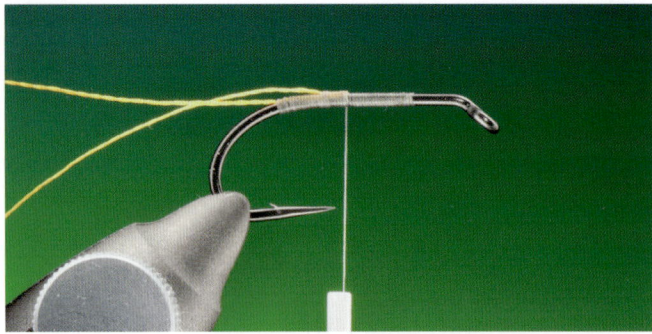

5 Fold over the neon thread and secure so the thread is doubled.

6 You will need a little scruffy dubbing for the body, here I'm using SLF Squirrel dubbing

7 Spin the dubbing onto your tying thread and start wrapping the body. Note that I have left 2-3mm of rib attached to the hook bend before I start the dubbed body.

8 The finished dubbed body should have a cigar shape. Finish quite a distance from the hook eye. If the dubbed body is too scruffy trim it down and to shape.

9 Take your neon tying thread and make five or six wraps to form a hot spot at the rear.

10 Continue with the neon ribbing and make five or six evenly spaced open turns. Tie off.

11 Select two similar-sized hen mallard breast feathers with nice markings. Look for two feathers that have square ends.

12 Strip off all the down material from the lower stem, and then the barbs, so they are a similar size.

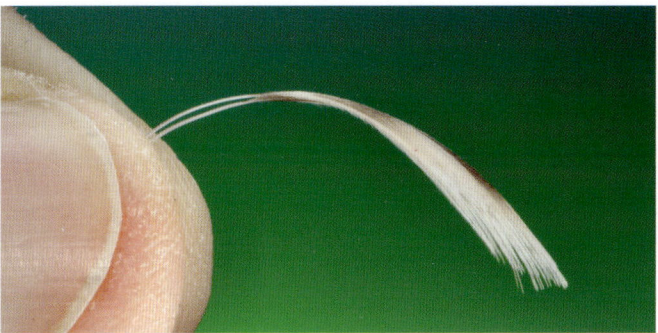

13 Place the feathers concave side down on top of each other. And match up the tips so they are level.

14 Carefully place the two feathers on top of the hook shank and fix with a couple of loose wraps of tying thread.

15 Take hold of the stems of both feathers, over the hook eye, and gently pull through the wraps of tying thread until the wings are the correct length. As you pull, you will see that the feathers work their way down each side of the body.

16 The finished wing should look something like this. Trim away the surplus and tie down the stems.

17 You will need to select a nice speckled partridge feather for the front hackle.

18 Prepare the partridge feather as shown.

19 Tie in the tip of the feather, and wrap as you would a traditional wet fly hackle, with the barbs pointing backwards.

20 Once you have wound your partridge hackle, tie it off and secure with a single whip-finish.

21 Remove the partridge hackle stem.

22 If you don't have Super Fine dubbing, you will need a dubbing that is fine and long in the fibres.

23 Spin the dubbing onto your tying thread and wrap a head as shown.

24 Finish with your tying thread behind the hook eye.

25 Make a couple of whip-finishes and remove your tying thread.

26 Using a soft toothbrush or dubbing brush, brush the long dubbing fibres, *backwards only!* Do this around the whole fly.

27 The long dubbing fibres will veil the hackle and wing, holding them back and down. Give the head a drop of varnish if you wish.

28 View of the diving caddis from below.

Dyret – The Animal

Deer hair tail and head in one • Dubbed dry fly body • Palmered body hackle

The Animal or Dyret as it's known in its native Norwegian, is a relatively new pattern, but one that has been embraced with open arms by Scandinavian flyfishers. Its Norwegian originator, Gunnar Bingen, developed this pattern for fishing both trout and grayling on the famous river Rena in Norway.

He's quoted as saying, 'it's really nothing but a cross between a super pupa and a devil bug'. But this offspring which arose from the cross-breeding of these two patterns has proved to be a deadly one.

Originally it was thought to imitate an emerging caddis, which it does so elegantly. But it really comes into its own by pressing all the right buttons for feeding trout when swimming caddis pupae are on the go, from late afternoon and into the night.

Night fishing with this pattern under a caddis hatch can put you on the verge of madness, listening for rises in the direction of your stripped fly and striking to sound instead of sight!

The deer hair used for this pattern should be from the winter coat as reasonably long hair makes this pattern easier to tie, giving you something to grab hold of when trimming the head. You can decide what type of head you prefer, whether a small tight trimmed one or a large open one that pushes more water as you tug it in. Some tyers prefer to use an extra large hackle so the fibres reach out further from each side. I feel the Dyret fishes best high and dry, so it should be well dressed with a floatant. I personally like to dress it first with a liquid floatant, give it a good shake, blow off the excess then follow with a quick shake in a powder floatant. This results in a super water-repellant dry fly.

On flowing water I feel it fishes best in smaller hook sizes using a dead drift method, making presentation over rising or feeding fish if possible. It is also very good for searching faster riffles and pocket water. Just letting the fly drift quickly through possible holding spots is extremely effective. I have also had great success with this pattern on still waters retrieving with short strips across the surface, with pauses at intervals while you just let the fly sit on the surface for a few seconds and then start again with short pulls. It's normally at this exact moment the fish will take explosively.

When wrapping the hackle, palmer style, don't make too many turns. If you wrap the hackle too tight along the whole body you may find that when you cast it, it spins like a helicopter and will twist your tippet into the mother of all tangles. This should be avoided at all costs, especially during night fishing!

Although the original Dyret retained the full hackle, most tyers now trim it on the underside as in this pattern.

Regarding colour, all the olives work well for me from light to dark but many swear by grey and even yellow bodies; for night fishing, black bodies or even entirely black flies are the trend. But try your own favourite trout and grayling combinations just as Gunnar Bingen did. You never know, you may be on to something!

TECHNIQUES MASTERED

Deer hair tail and head in one
- An easy technique that lets you tie in a single bunch of deer hair for a tail head and underbody.

Dubbed dry fly body
- How to apply dry fly dubbing over the deer hair underbody, without crushing the deer hair.

Palmered body hackle
- Wrapping a palmered body hackle from tail to head and trimming it for a lower profile.

Tying the Dyret – The Animal

THE DRESSING

Hook: Mustad Heritage R30AP # 10
Tying thread: Sheer 14/0 tan
Tail and head: Natural deer hair
Hackle: Brown
Body: Super Fine dubbing olive brown

WATCH THE VIDEO

youtube.com/watch?v=4VjZy-un5qs

Tying the Dyret – The Animal with Barry Ord Clarke

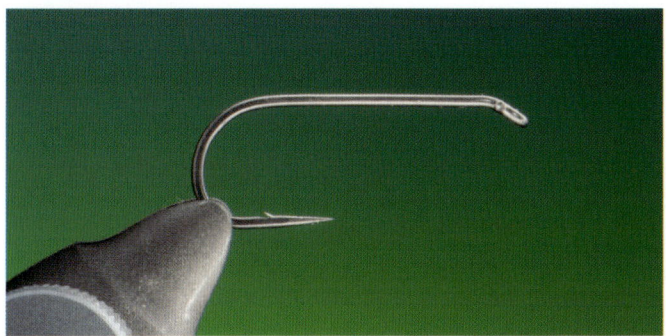

1 Secure your dry fly hook in the vice. Make sure that the hook shaft is horizontal. If you have a true rotary vice, centre the hook.

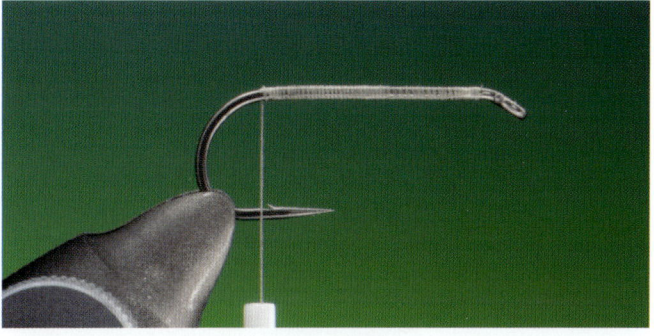

2 Attach your tying thread as shown, a little behind the hook eye, and run a foundation the whole way along the hook shank, to the tail base.

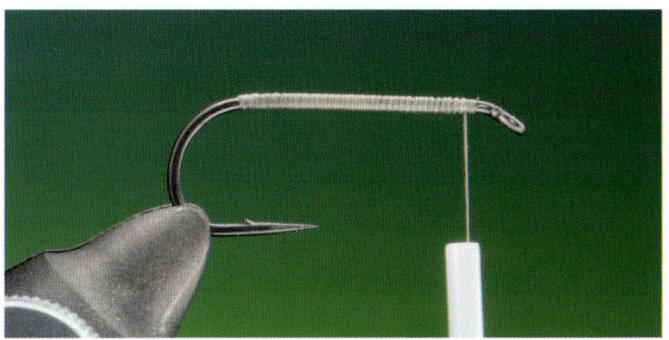

3 Run your tying thread back up the hook shank and finish behind the hook eye.

4 Cut a small bunch of long deer hair, with nice marked unbroken tips. Long deer hair makes it easier to handle when tying in.

5 Clean the hair with a comb so that all the underfur and shorter hairs are removed. Place the hair in a stacker and even up the tips.

6 Once the hair is stacked, measure the tail length. Keeping hold of the bunch so the tail is still the correct length, spin your bobbin anti-clockwise, so it attains a flat profile then tie down the head behind the hook eye.

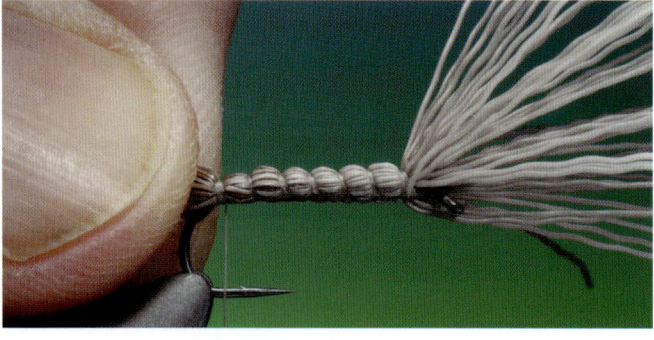

7 While still holding the bunch of deer hair, slide your grip back while you wrap as shown, until you reach the tail base. *Take care not to wrap too hard.*

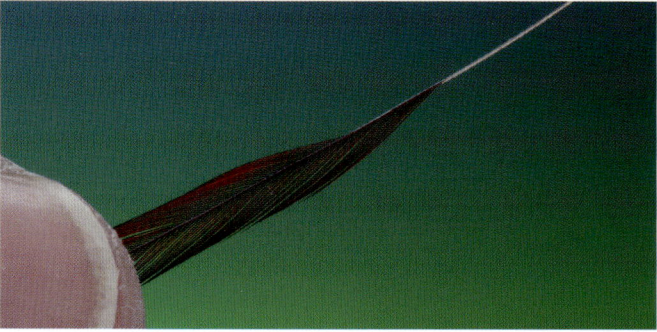

8 Prepare a saddle hackle by stripping the lower stem as shown here, removing a few more barbules from the leading edge of the stem than the other.

9 Attach the hackle stem at the rear, leaving a millimetre or two between the tail and the hackle.

10 Fix the rest of the hackle stem along the whole length of the hook shank.

11 Now run your tying thread back to the hackle.

12 Choose your colour for the body. Here I'm using olive brown Super Fine dubbing.

13 Spin a little dubbing on your thread, not too tight. Catch in the dubbing with a single turn between the tail and the hackle.

14 Spin the rest of your dubbing, loosely – it shouldn't compact the deer hair underbody but should be loose enough so the palmered body hackle sits a little down in it. Wrap the dubbing rope forward until you reach the head.

15 Attach a hackle plier to your hackle and carefully wrap the hackle forward in even open turns. Take care that you don't wind the hackle too densely. Tie off and cut away the excess hackle.

16 Taking hold of the whole deer hair bunch with your left hand, fold it back over the body. Wrap your tying thread forward behind the hook eye and secure with a couple of whip-finishes.

17 Once secure, remove your tying thread.

18 Using sharp serrated scissors, trim down the deer hair head. I like to leave a slightly larger head so that it pushes a bit more water when retrieved.

19 If you would like the Dyret to fish with a lower profile, rotate your vice and trim away the hackle barbules on the underside of the fly.

20 I prefer to leave the trimmed hackle on the underside a little stubbily, as you can see here.

21 The underside view of the finished Dyret.

17

Welshman's Button

Twist & wrap CDC body • Mallard feather caddis wings
• Hackle stem antennae

This little caddis fly is my interpretation for the Welshman's Button, or Chocolate Drop as it is also known, and it has been the demise of many a river trout.

Despite its semi-realistic appearance and the somewhat unfamiliar techniques used, this is a remarkably simple pattern to master. Once again, I can't stress how important fly proportions are! So please pay constant attention to the proportions throughout each step.

The CDC twist and wrap abdomen technique is one from Marc Petitjean that I have embraced. This ingenious method of using a single CDC hackle to create buoyant segmented bodies can be adopted for many dry fly patterns. But there is a specific procedure to follow when using this technique.

The CDC hackle required needs to be long, preferably domestic duck or goose CDC. Petitjean's own CDC is perfect for this.

Once you have secured the CDC hackle to the rear of the hook shank, you must attach a good hackle plier that has an iron grip. Make sure that the hackle plier is in line with the CDC stem. Before you begin to wrap the hackle, twist it twice using your hackle plier. As you do this you should see the CDC barbs closest to the hook shank wrap together into a cord. Keeping tension on your hackle plier, make a couple of close touching wraps. It's important that you pull the CDC fibres into the stem each time you twist. Twist your hackle plier again and make a couple more wraps. Continue until the whole body is complete and tie off.

It's important that you twist slightly after every wrap. If you try and twist the whole hackle at once, it will break!

If you require twist and wrap bodies for smaller flies, strip the barbs off one side of the CDC hackle before you tie it in. If you need bodies for larger hook sizes, you can use two CDC hackles together. You can also use two different coloured CDC hackles for a bi-coloured body effect.

For the wings you will need two mallard drake chestnut brown neck feathers with white tips, *see step 13*. These feathers are not generally available commercially. Tyers who are wild fowlers themselves, or who are acquainted with shooters, should try to get their hands on a whole skin and experiment with the entire plumage of this most useful duck.

TECHNIQUES MASTERED

Twist and wrap CDC body
- A Marc Petitjean technique for CDC segmented dryfly bodies that works for most hook sizes for trout.

Mallard feather caddis wings
- A nice Devaux-style caddis wing technique, using two chestnut brown mallard drake neck feathers.

Hackle stem antennae
- Styling two fine antennae from stripped dun cock hackle stems tips – and the correct way to attach them.

Tying the Welshman's Button

THE DRESSING

Hook: Mustad Heritage R50 # 12
Tying thread: Sheer 14/0 brown
Body: Chocolate brown CDC hackle
Wings: Two chestnut brown Mallard drake neck feathers
Antennae: Two dun cock hackle stems
Hackle: Chocolate dun cock hackle

WATCH THE VIDEO
youtube.com/watch?v=dzXmxNPdFSw&t=334s

Tying the Welshman's Button with Barry Ord Clarke

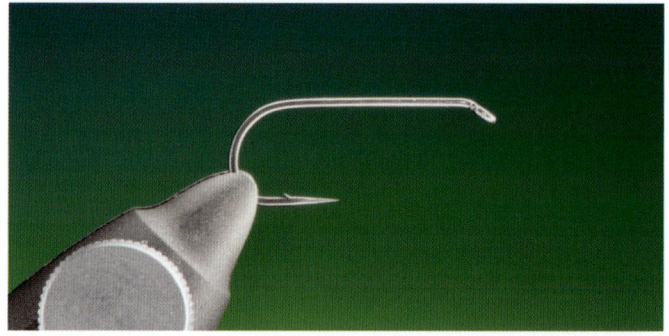

1 Secure your hook in the vice. Make sure that the hook shaft is horizontal. If you have a true rotary vice, centre the hook.

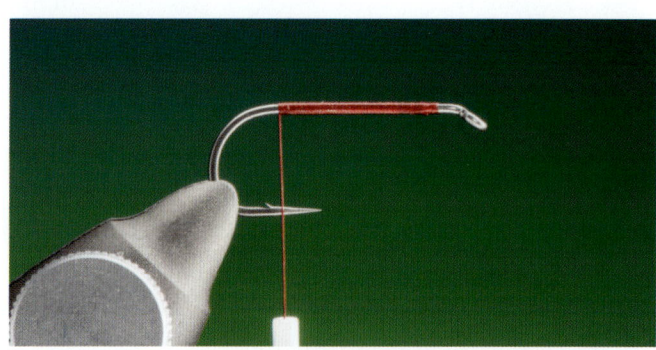

2 Attach your tying thread as shown, a little behind the hook eye and run a foundation along the whole of the hook shank.

3 You will need a long chocolate brown CDC hackle.

4 At the very rear of the tying thread foundation, secure the CDC hackle as shown with a couple or three loose turns of tying thread.

5 Keeping a little tension on your tying thread, carefully pull through the CDC hackle until the tip fibres are behind the hook eye.

6 Once positioned, you can secure the CDC tip by wrapping your tying thread forward to the hook eye.

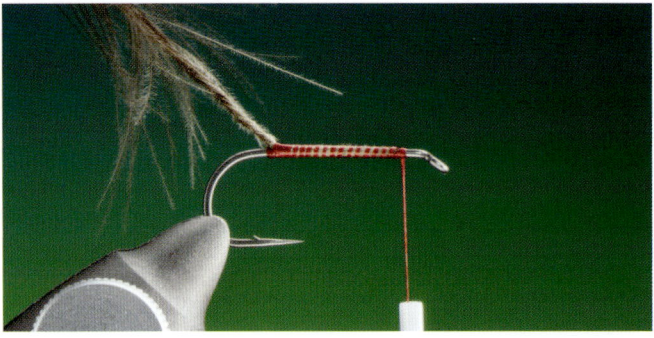

7 Attach your hackle plier to the CDC hackle and twist the hackle twice.

8 Now make a couple of wraps around the hook shank and twist the hackle again. Repeat until you have the desired body.

9 Tie off the hackle and cut away the excess hackle stem.

10 Using fine pointed scissors trim off the body fibres. This should result in a nice slender CDC body.

11 You will now need two chestnut brown mallard drake neck feathers with white tips.

12 Prepare the two feathers by stripping off the fibres until both feathers are the same size.

13 Position the two feathers on top of each other and with a couple of loose turns of thin thread attach them on top of the hook shank as shown.

14 Now you can carefully manoeuvre the wing into the correct position by slowly pulling on the feather's stems.

15 Once you are happy with the wing, secure with a few more tight wraps of tying thread and then trim away the excess stems. Pay special attention to the wing length and position.

16 Select two dun cock hackles of a similar size.

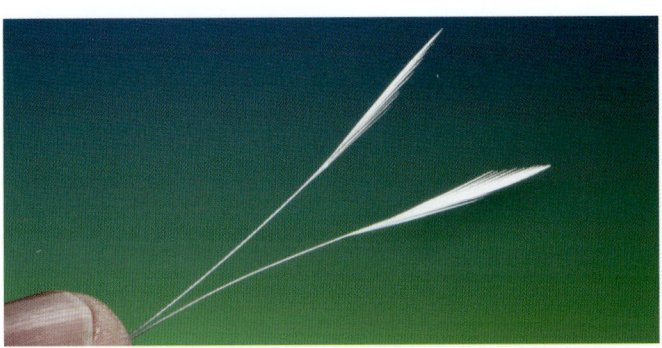

17 Leaving only the tip of the feather, strip away the remaining fibres.

18 Place both hackle stems on top of the hook shank with the fine tips facing forward and secure with a few wraps of tying thread.

19 Once secure, lift the rear of the stems 90 degrees to the hook shank. This will make them easier to cut away.

20 Trim off the rear stems as close to the wing as possible.

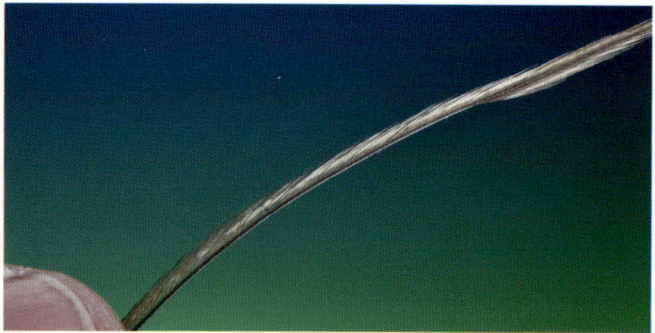

21 Prepare a chocolate dun cock hackle by stripping off the barbs from the leading edge of the hackle.

22 Tie in the hackle close to the wing, so it stands 90 degrees from the hook shank.

23 Trim off the hackle stem to the correct length for tying down. I prefer to trim it at an angle, so it tapers down to the hook eye.

24 Now tie down the hackle stem and finish with the tying thread behind the hook eye, but under the antennae.

25 Attach your hackle plier and carefully wrap the hackle in tight touching turns forward.

26 Tie off the hackle and trim away the excess.

27 Make a couple of whip-finishes and remove your tying thread.

28 Carefully place a tiny drop or two of varnish on the head, and the Welshman's Button is finished.

29 An aerial view of the fly showing the correct proportions and form.

Fluttering Caddis

Segmented abdomen technique • Double Wally wings • Mallard flank antenna

I have chosen to include this pattern in this book, more for its tying techniques than its fishing merits. Although you could fish it, you may feel that it's a little too time-consuming to tie and then lose to a fish or a tree!

The flyfishing club where I am a member has a motto, 'pescator non solum pescatur' – 'there is more to fishing, than catching fish'. A truer thing could never have been said about flytying either.

Two mallard flank feathers, some dubbing and some practice, are all that are required to make this little fluttering caddis fly.

Your choice of winging material for tying Wally wings is paramount, if you wish to succeed! If you have purchased a packet with mixed mallard or teal flank, only a limited amount of these will be useable. Here I can recommend purchasing a whole mallard drake skin. Having a whole skin has many advantages. The price of a whole skin is nominal, when you think what you pay for a little packet stuffed with feathers that contains only a couple of grams. With the whole mallard drake skin, all the feathers are perfectly packaged by nature, keeping them all in the same direction and neatly stacked on top of each other. This also makes selection of individual feathers in the size required easier. If you feel the investment of a whole skin is beyond your budget, consider purchasing one as a collective with other flytying friends and splitting it.

There is a mixed school of thought in the flytying and fishing fraternity when it comes to fishing with Wally winged patterns. Some say they are hopeless and constantly twist the leader under casting while others, like my good friend and world class flytyer Trevor Jones, who has been one of the pioneers behind this pattern, swears by them. No matter what your opinion may be, I think we can all agree that they make a beautiful-looking fly and are fun to tie.

Regarding the technique used here that I developed to tie Wally wings, you will require a small plastic tube to hold the barbs in place for tying in. I have experimented with several types of tube for executing this technique and have found the absolute best is a tapered reserve nozzle from a small bottle of UV resin. Depending on what size of Wally wings you intend to tie, you will firstly have to adjust the tip opening of the nozzle to the correct size of feather to be used. This is simply achieved by preparing a flank feather as in step 15, and checking if you can pass it all the way through the tube. You don't want to tie the wings in and not be able to remove the tube because the feather stem is too thick. Cut a couple of millimetres from the end of the tube and try again. Repeat until the correct nozzle opening is attained.

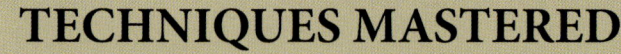

TECHNIQUES MASTERED

Segmented abdomen technique
- How to use regular Super Fine dubbing to make a segmented abdomen effect.

Double Wally wings
- Making and mounting double mallard flank Wally wings for a spent caddis fly.

Mallard flank antenna
- Attaching caddis fly antenna.

Tying the Fluttering Caddis

THE DRESSING

Hook: Mustad Heritage R43 # 12-14
Tying thread: Sheer 14/0 olive
Abdomen: Olive Super Fine dubbing
Wings: Two mallard flank feathers
Thorax: Brown Super Fine dubbing
Antenna: Two mallard flank barbs

WATCH THE VIDEO

youtube.com/watch?v=jqlT5Ltoe0E

Tying the Fluttering Caddis with Barry Ord Clarke

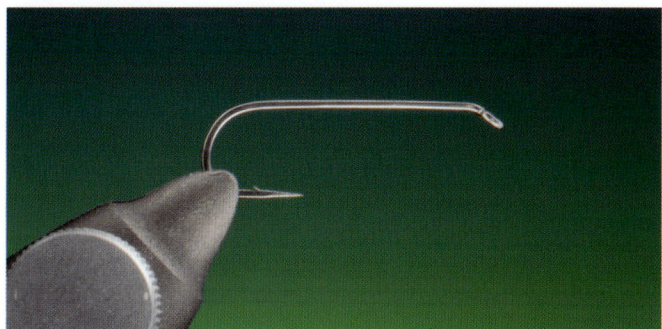

1. Secure your dry fly hook in the vice. Make sure that the hook shaft is horizontal. If you have a true rotary vice, centre the hook.

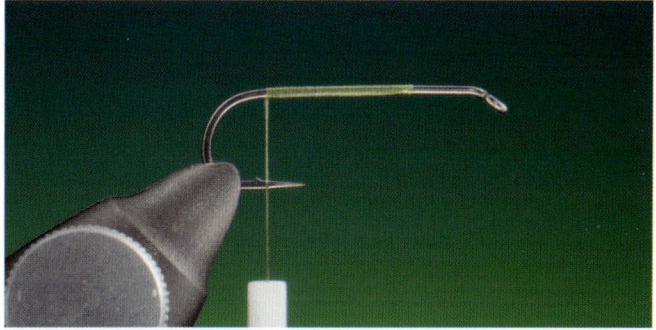

2. Attach your tying thread as shown, a little behind the hook eye, and run a foundation along the hook shank.

3. You will now need some olive dry fly dubbing. I can recommend Super Fine if you haven't used it before. It floats well and is extremely easy to dub.

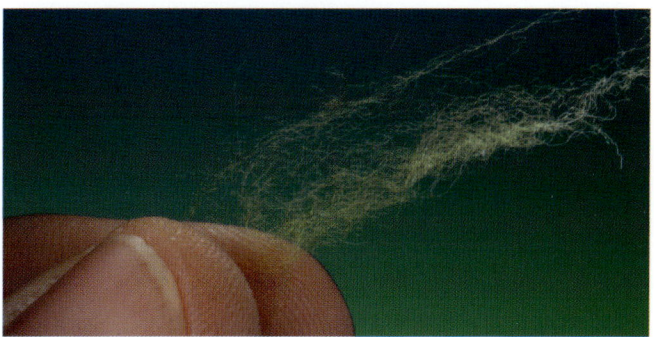

4. When pulling the dubbing from the box, do so very gently, so you only get a wisp of dubbing. This will help to keep it fine and even when applied to the tying thread.

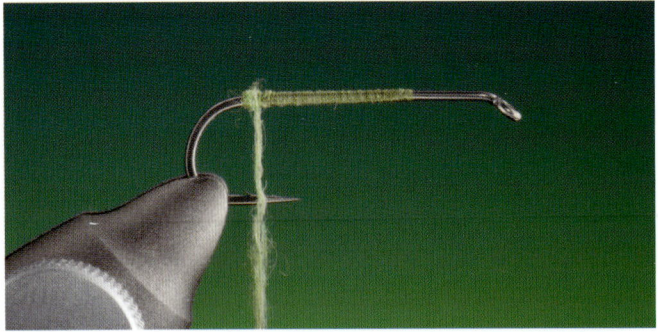

5 Take the finest end of your dubbing wisp and roll it between your finger and thumb onto the upper section of your tying thread. Once secure, slide it up to the hook shank and make a couple of wraps to catch the dubbing in.

6 You can now spin the remaining dubbing onto the tying thread, increasing the amount as you go so you get a tapered abdomen. If you have a true rotary vice, use it to wrap the dubbing.

7 Once you have dubbed approximately two thirds of the hook shank, tie off the dubbing as shown on top of the hook shank. Don't remove the excess dubbing!

8 Before you start on the wings you will need a small plastic tube. This one is from a bottle of UV resin. You will have to cut a couple of millimetres from the tip of the tube in order to make the opening larger.

9 Select a mallard flank feather with long barbs on each side of the rachis and nice markings. Make sure that there are no short barbs in the two sections to be used for the wings.

10 Wet the barbs at the tip of the flank feather and pull to a point. Draw back the lower barbs to form two wings, as shown.

11 Take hold of the tip of the flank feather and and cut away the thicker lower half of the flank feather stem.

12 Place the tip of the prepared flank feather into the tube and carefully pull through until you have the size of wings required.

13 Take a little brown Super Fine dubbing and dub a small section of the thorax. You will now need to wax your tying thread.

14 Now, keeping the mallard flank in the tube, offer the wings up to the correct position on the thorax and secure with a couple of turns of tying thread. While retaining tension on your bobbin, carefully position the wings into their final resting place.

15 Once in position, secure with a few more turns of thread. Still retaining tension on your bobbin, carefully slide off the tube.

16 You can now cut away the remaining stem of the mallard flank feather and tidy up with a few more turns of tying thread.

17 While holding the tip of the mallard flank in one hand, carefully separate the first two barbs from the tip with the other hand. You should take care here to select only two barbs from each side – no more, no less!

18 Now you're ready to split the wings. While still holding the tip of the flank feather, take hold of the two barbs, carefully tear them away from the rachis (central feather shaft) as shown, until it tears all the way down to the thorax.

19 You can now repeat step 18 to make the second wing.

20 With a pair of fine-pointed sharp scissors, carefully cut away the centre rachis and the remaining barbs on the wing tips.

21 Once you are happy with the wings and position, fold over the dubbing wisp and tie down as shown. This will form the first section of the thorax and hold the wings in position.

22 Fold back the dubbing wisp and make another small dot with brown Super Fine dubbing.

23 Prepare another mallard flank for the slightly larger front wings.

24 Tie in the front wings as with the first, taking care that these are larger than the first.

25 With a pair of fine-pointed sharp scissors, carefully cut away the centre rachis and the remaining barbs on the wing tips.

26 Take one of the flank feather tips you had as surplus from the wings.

27 Select two barbs from the tip and tie in as shown for the caddis fly antenna. Take a little more brown Super Fine dubbing and dub the remaining thorax.

28 Fold over the remaining dubbing wisp and tie down tight into the hook eye.

29 Carefully trim away the surplus dubbing wisp and build a small, neat head with a couple of whip-finishes. Remove your tying thread.

30 Side view of the Fluttering Caddis.

163

31 View from below.

32 The finished Wally wing Fluttering Caddis.

Red Panama

Golden pheasant tippet tail • Palmered body hackle
• Double hackle tip wings • Speckled partridge front hackle

There are few surface patterns that have so many procedures and techniques to wade through, as this striking fancy French dry fly. But don't be fooled by its good looks, the Panama has a criminal record!

The Panama, or Panama Palmer as it was also called, is one of the oldest French artificial patterns. Its origin is often attributed, rightly or wrongly, to journalist and fishing writer Tony Burnand. It can be traced back to the interwar period: one of its earliest references in literature, dates back to 1935 (*Au Bord de l'Eau* magazine), but we know that it had been used extensively and successfully for years, notably on Normandy chalkstreams and the rivers of Brittany.

Like many fancy flies, the Panama doesn't imitate anything in particular, but its considerable proportions (it can easily be tied on a size 8 hook) and silhouette make it a very effective pattern during hatches of the larger mayflies. Tony Burnand, Léonce de Boisset and Charles Ritz used it extensively, and mention it in their books and articles. It was mass-produced and marketed from the 1930s onwards. It became such a popular pattern during this period that the Ragot company had the Panama tied in bulk by convicts in French prisons.

With the development and introduction of finer tippet materials, the Panama, along with other large fancy patterns of the era, became unfashionable as did dry fly fishing in France. It should be added that, being a large bushy fly, fished on a fine tippet, it has a tendency to helicopter.

When preparing and tying in the golden pheasant tippet, in order to get a nice, even, balanced tail, with straight black barring, you can attain a much better result if you keep the barbs on the hackle stem as shown in Step 11. You use only one side of the barbed V for the tail. This technique allows you to handle and position the tippets, without losing the natural alignment of the markings. Once you have secured the tail, just trim away the excess from the side used, and you have one side remaining for the next Panama tail! The remaining centre you cut away from the V can be stored away for another time. No waste and perfect tails.

The original Panama had a body of natural beige raffia and although this was buoyant, it made for a uneven and bulky body.

The red Panama was a popular variant which replaced the raffia body of the original with one of red floss silk.

Other than being an absolutely stunning-looking pattern which is a joy to tie, we have several techniques that can be practised and mastered when tying the red Panama. The upright twin wings, golden pheasant tippet tail, palmered body hackle, floss silk body… All the techniques required here are essential for the all-round consummate flytyer.

TECHNIQUES MASTERED

Golden pheasant tippet tail
- An easy technique for achieving perfect golden pheasant tippet tails every time.

Palmered body hackle
- How to prepare and attach a forward-palmered dry fly body hackle.

Double hackle tip wings
- Traditional vertical dry fly winging technique for attaching and raising of double wings.

Tying the Red Panama

THE DRESSING

Hook: Mustad Heritage R43 # 12

Tying thread: Sheer 14/0 black

Tail: Golden pheasant tippet

Butt: Black tying thread

Body: Red silk floss

Body hackle: Natural brown saddle hackle

Wings: Two fine brown and two grizzle hackle tips

Front hackle: Speckled partridge shoulder hackle

WATCH THE VIDEO

youtube.com/watch?v=_W7_HLRy2zM

Tying the Red Panama with Barry Ord Clarke

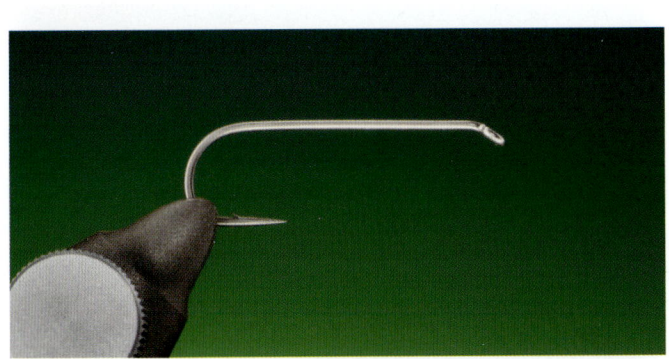

1. Secure your hook in the vice. Make sure that the hook shaft is horizontal. If you have a true rotary vice, centre the hook.

2. Attach your tying thread as shown, a little behind the hook eye and run a foundation a short way along the hook shank.

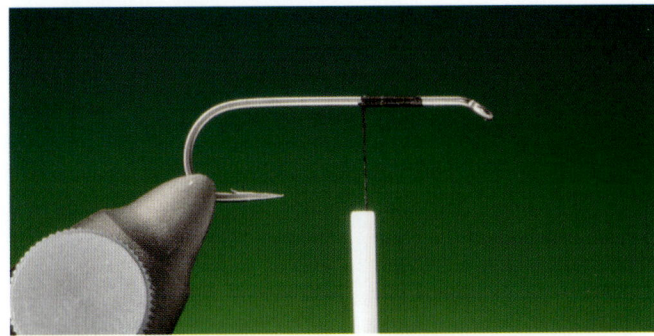

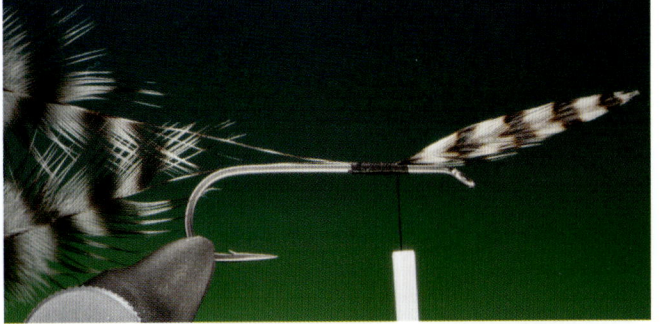

3. Select two fine grizzle hackles and strip away the barbs from the lower section, retaining a little of the web material at the base of the hackle. This makes them easier to handle. These two wings should be approximately the same length as the hook shank.

4. Holding both wings with your left hand, align the tips and position them on top of the hook shank, with about quarter of the hook shank's length from the hook eye.

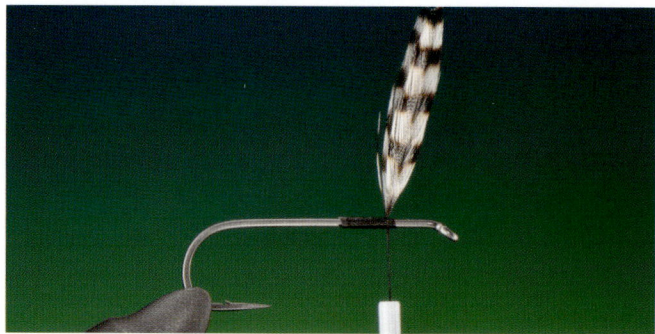

5 Once you have secured the wings, trim away the surplus hackle stems. You can now lift the wings vertically and make a few wraps of tying thread, tight in front of the wings to keep them vertical.

6 Make a couple of figure-of-eight wraps with tying thread between and around the two wings, so they are secured slightly forward at 45 degrees from the hook shank.

7 Now select two fine brown cock hackles and prepare as before. These should be a few millimetres longer than the grizzle wings.

8 Tie in the brown wings as shown on the inside of the grizzle wings.

9 Once again, slightly forward and at a 45 degree angle.

10 You will now need a natural golden pheasant tippet feather with nice straight dark bars.

11 Using sharp fine-pointed scissors, carefully trim away the centre of the tippet feather as shown, retaining about 10-12 barbs on each side.

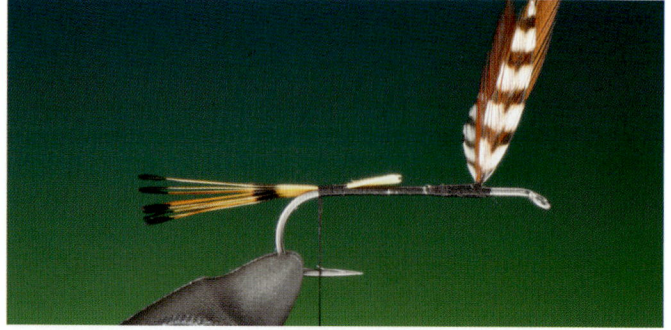

12 Keep the barbs on the stem secure with a few wraps of tying thread. Take care that both black bars on the tippet barbs are visible as shown.

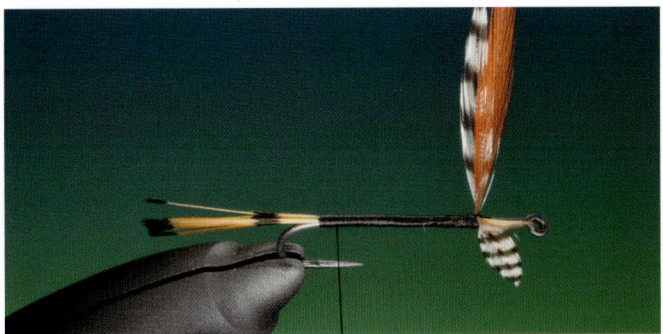

13 Once the tail is secure, run a neat foundation of tying thread along the whole hook shank, making the underbody as even as possible. Take care that you have a nice black butt 2-3mm at the tail base as shown.

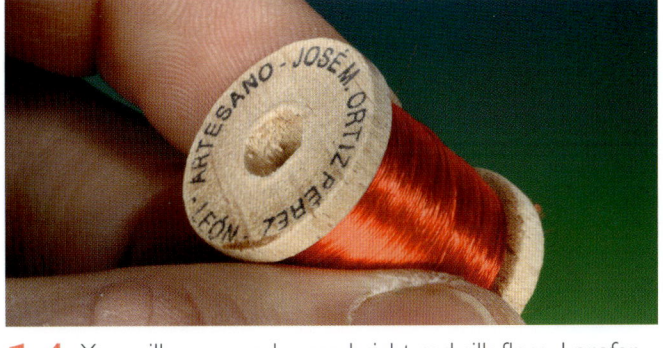

14 You will now need some bright red silk floss. I prefer to use real silk floss, rather than synthetic. It's a little more expensive but it's easier to use and the results are always better.

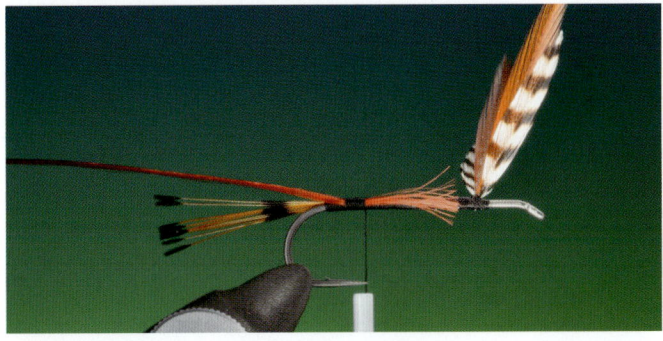

15 Cut a short length of red silk floss and tie this in, 2-3mm from the tail base. Take care that the short end of the floss is long enough to cover the whole length of the hook shank. This will give you an even body when wrapped.

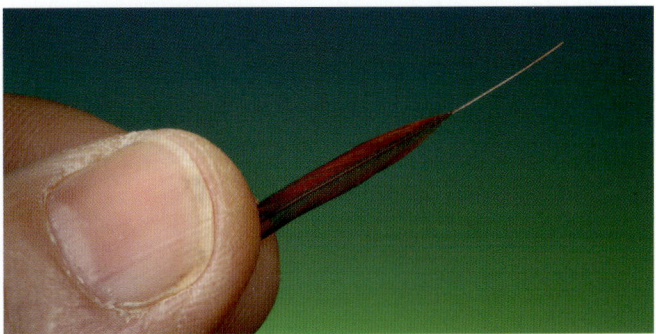

16 Prepare a natural brown saddle hackle as shown.

17 Tie in the hackle as shown, slightly forward of the red silk floss. Tie down the silk floss forward over the whole hook shank, taking care to keep the body even over the whole hook shank.

18 Trim the hackle stem to the correct length for the hook shank and secure. Finish with your tying thread in front of the wings.

19 Firstly make a couple of turns of silk floss behind the hackle stem, then wrap the silk floss forward in even, slightly overlapping turns, covering the whole body. Finish at the rear of the wings and tie off.

20 Attach a hackle plier to the body hackle and wrap over the whole body in even open wraps. When you get to rear of the wings, tighten the wraps as with a traditional collar hackle, finishing with a few wraps forward of the wings, leaving enough space for the partridge hackle.

21 Select a nice speckled partridge shoulder feather with barbs a little longer than the body hackle.

22 Strip away the lower web fibres and then prepare as shown.

23 With only a couple of wraps of tying thread, position the partridge hackle.

24 Once in position, trim away the surplus and secure. Attach a hackle plier and wrap the partridge hackle forward. Tie off.

25 Trim away the surplus partridge hackle and wrap a small neat head.

26 Make a couple of whip-finishes and remove your tying thread. Finish by giving the head a couple of coats of clear hard varnish.

27 Front view of the finished Red Panama.

Giant Stone Fly

Realistic Stone Fly bodies • CDC wings • CDC dubbing brush thorax

There are hundreds of patterns that are designed to imitate one of the largest trout foods to be found, *Pteronarcys californica,* the famous salmon fly or giant stone, so why another one here?

Stoneflies spend most of their lives as nymphs, clinging to the bottom of rocks in well-oxygenated riffles. When the the time is right, the large alien-like nymphs leave the safety of the river bed where they have lived from one to three years and crawl out of the water onto the bank-side stones and vegetation where they proceed to emerge into the winged adult. These are huge in size, up to 3 inches long and extremely poor aviators so they provide the flyfisher with some of the most exciting surface sport there is to be found.

We flytyers are always looking for new materials, techniques and how to put them to good use. This pattern uses an innovative little extended body pin that you mount in the vice jaws to build on the stonefly body. These little extended body pins were designed and made by Dutch flytyer Maarten van Eijk and come in a set that includes three sizes, small, medium and large. If you are handy, you could perhaps make your own!

With the use of these extended body pins you are able to make perfect realistic foam stonefly bodies that are robust and will float all day long and then some! All you need are a few basic materials which I am sure anyone reading this already has to hand. With a little creativity, you can change the colour of foam, tail, thorax and wing materials, to create your own patterns that are guaranteed fish-catchers.

The extended body pins are available from: https://www.maartenvaneijk.nl/product/body-pins/

TECHNIQUES MASTERED

Realistic Stone Fly bodies
- How to fashion realistic foam and deer hair extended stonefly bodies.

CDC wings
- Making large stonefly wings from two CDC feathers coated with a little head cement.

CDC dubbing brush thorax
- How to utilise a standard natural CDC dubbing brush to make combined high floating thorax and legs for the stonefly.

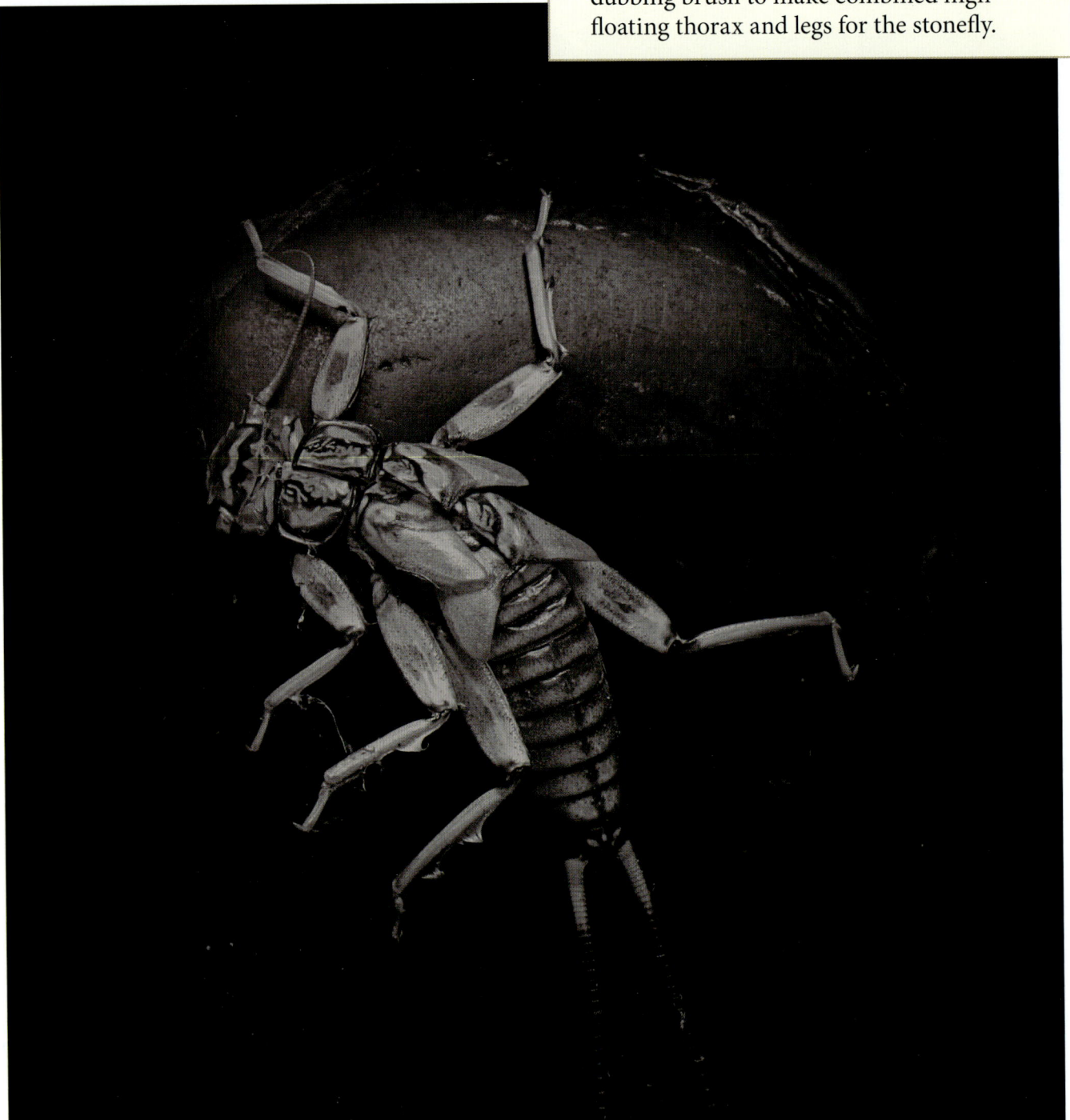

Tying the Giant Stone Fly

THE DRESSING

Hook: Mustad Heritage C49 SAP # 10
Thread: Sheer 14/0 brown
Underbody: Natural deer hair
Tails & Antenna: Moose body hair
Overbody: Black 1mm foam sheet
Wings: Two large natural CDC feathers
Thorax/legs: Natural CDC

WATCH THE VIDEO

youtube.com/watch?v=t7c2JdpVHjc&t=19s

Tying the Giant Stone Fly with Barry Ord Clarke

1 The extended body pins come in a set of three sizes, S, M, L.

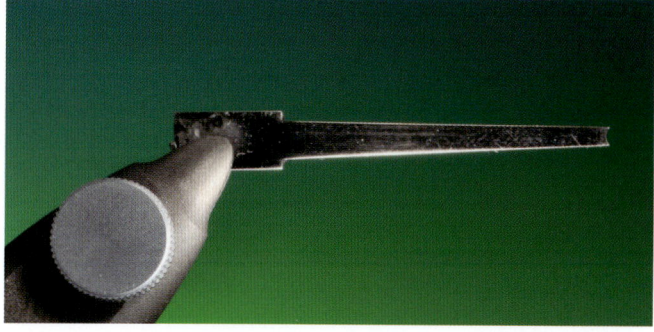

2 Mount the pin in the jaws of your vice as shown. Here I am using the medium size body pin.

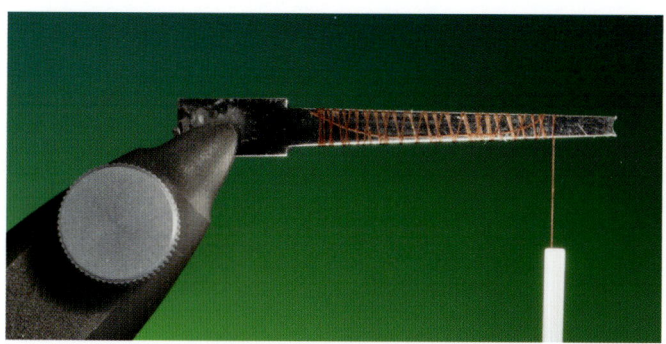

3 Attach your tying thread to the body pin and run a fine foundation over the whole length of the pin, finishing at the tail end.

4 Cut a small bunch of longish natural deer hair. Clean and stack to even up the tips.

5 You can now rotate your vice head so the pin is lying in the correct body position. Tie the bunch along the top of the body pin as shown. Care should be taken that the hair remains on top of the pin and not on the sides or underside.

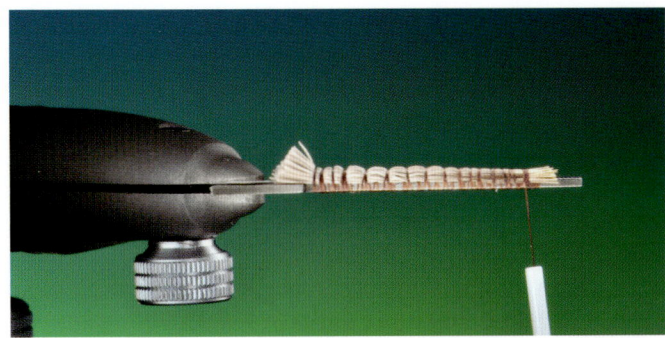

6 Using sharp scissors, trim off the the deer hair at both ends of the pin.

7 For the tails and antenna you will need some moose body hair. Select two straight black hairs for the tails.

8 Tie in the moose body hair tails, one each side of the body.

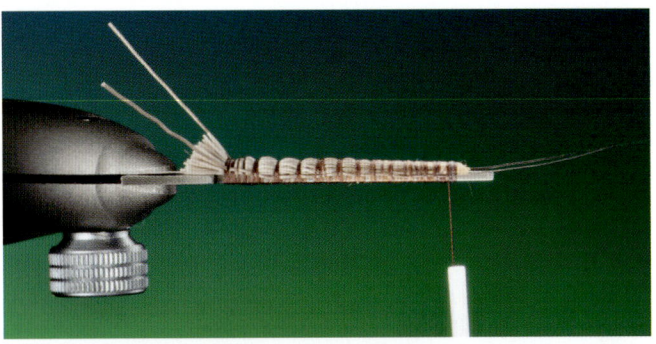

9 Secure the two hairs all the way along the body pin and back again.

10 Cut a 5mm strip of fine black foam. If your foam is thicker than 1mm, a good tip is to place your foam between two sheets of baking paper and iron it with your iron set to the lowest heat. This will compact the foam into a thinner sheet.

11 Once you have cut your foam strip, fold it so one side is a little longer than the other. The longer side will be used to create the thorax and head later.

12 Once folded, turn the foam and hold the folded end. While holding the foam securely, make two small diagonal cuts, taking care not to cut the foam in two.

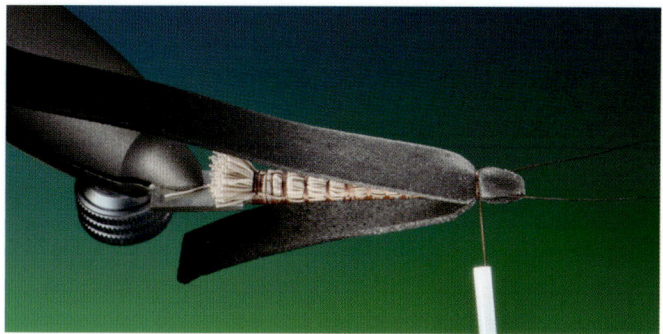

13 With your tying thread situated at the tail of the body place the foam strip, in between the tails with the short end of the foam on the underside of the pin. You can now make the first body segment with three or four wraps of tying thread.

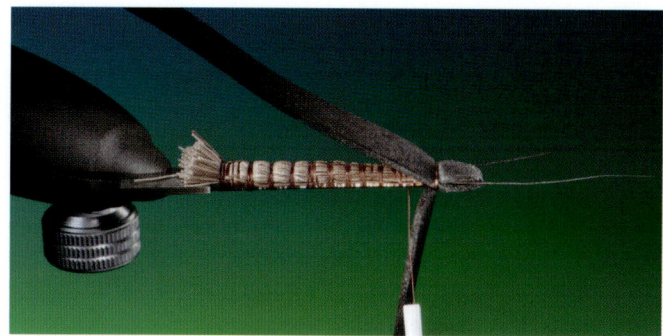

14 Once the first segment is secure, move your tying thread back onto the pin as shown, a little further along the pin for the next body segment.

15 You can now make the second body segment.

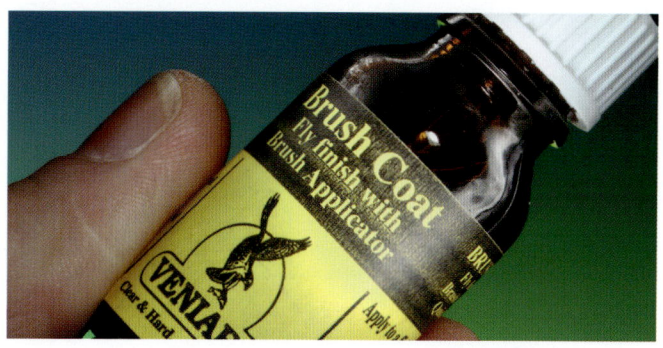

16 You will next need a little head cement to strengthen the body. I use Brush Coat from Veniard which is easy to apply and dries quickly.

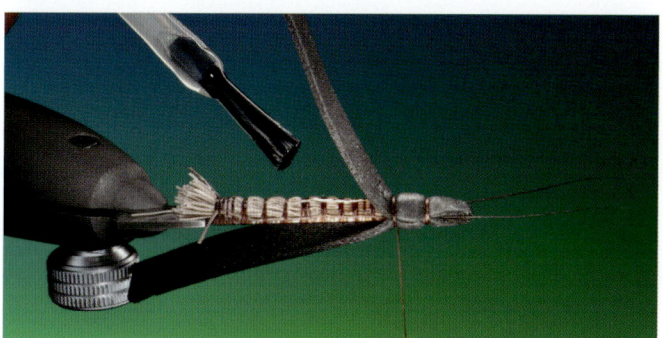

17 Apply a little head cement along the length of the deer hair underbody.

18 Keeping the wraps of thread firm but not too tight, continue up over the pin as before, until the full body length is segmented.

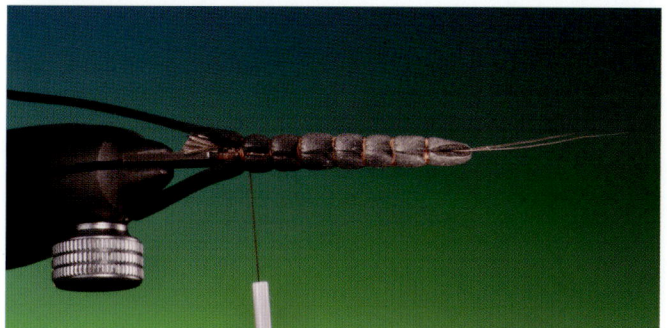

19 Here's a side view of the segmented body.

20 You can now make a couple of whip-finishes around the last body segment and remove your tying thread. Place a small drop of head cement on the whip-finish.

21 Taking hold of the rear of the body, gently push the body off the pin.

22 Once the body is off the pin you can roll it between your finger and thumb to create the correct body profile.

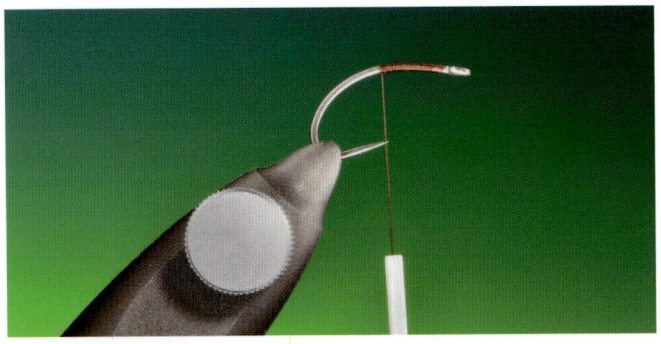

23 Place your hook in the vice with the hook shank horizontal and attach your tying thread.

24 You will need two more moose body hairs.

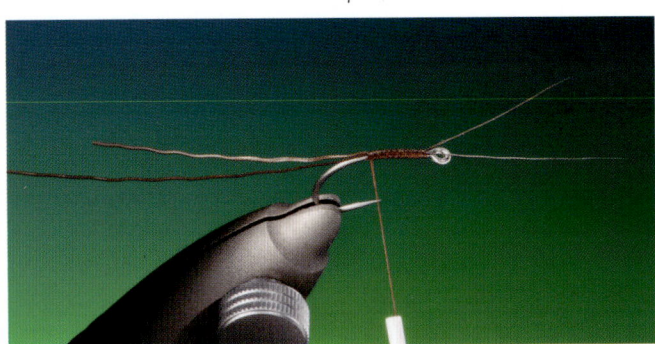

25 Tie in two more moose body hairs for the Stone Fly's antenna, over the hook eye.

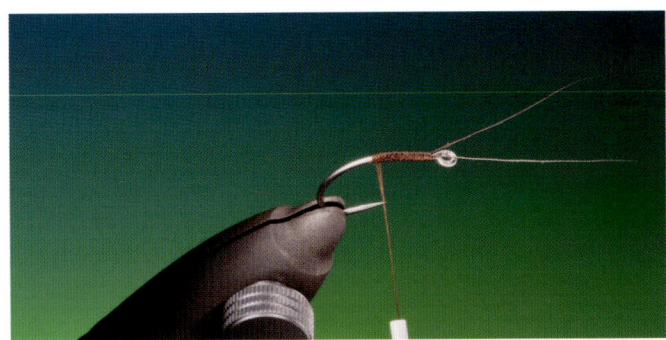

26 Trim away the butt ends of the moose hair.

27 Using the first segment, tie on your extended body with the remaining foam strip on top.

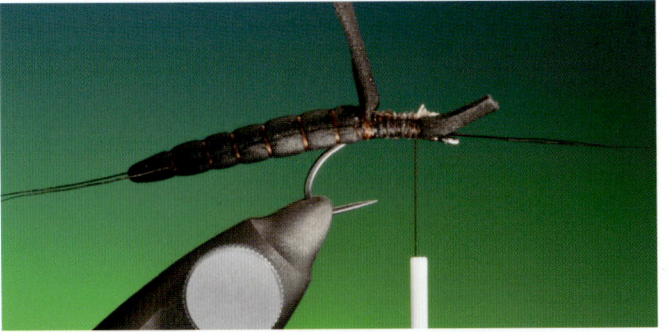

28 Secure the extended body as shown, taking care that it is central and balanced on the hook.

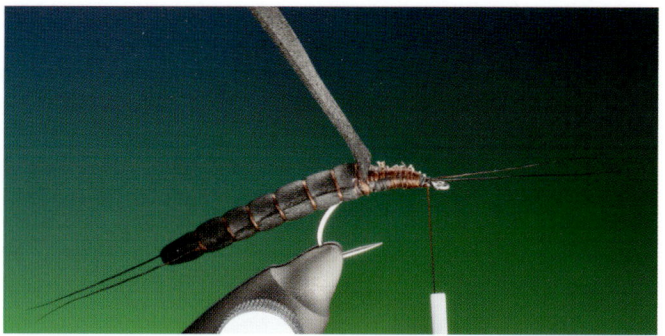

29 Once secure, trim away the shortest foam piece.

30 Select two long natural CDC feathers.

31 Brush a little head cement along the edges of each CDC feather and pull them between your finger and thumb to form the wings.

32 Now wrap your tying thread behind the remaining foam strip.

33 Load a magic clip with a single natural CDC feather.

34 Spilt your tying thread and spin the CDC into a dubbing brush. Make one or two turns of the CDC dubbing brush over the first body segment.

35 You can now tie in the first CDC wing at a slight angle, a little longer than the body.

36 Repeat with the second wing over the first, slightly split.

37 Trim off the CDC stems on the wings, and secure with a few wraps of tying thread.

38 Fold back the foam strip and secure with a couple of turns of tying thread.

39 Load another magic clip with CDC and make another dubbing brush.

40 Wrap the dubbing brush forward.

41 Take care that you cover the whole thorax, finishing just behind the hook eye.

42 Take hold of the foam strip and fold it over the thorax, taking care to keep it central. Tie this down with a few wraps of flattened tying thread.

43 Once secure, fold the strip back over the thorax and make a few wraps to form the head of the adult stonefly.

44 Trim off the remaining foam strip, as shown and make a couple of whip-finishes.

45 Remove your tying thread and the Giant Stone Fly is finished. Seen from above.

46 Side view with the extended body doing what it should.

47 If you use wing burners, you can make synthetic wings.

48 An alternative Stone Fly, this time with synthetic wings.

㉑

Braided Blue Damsel

Braided extended body • Foam para post • Ice dubbing thorax

Throughout the warmer months, these beautiful blue flies can be seen on most bodies of water, both still and flowing and in the course of the summer, they can be so prolific that they can prompt extremely selective feeding amongst trout. It's under such circumstances that it's astute to have a pattern that you know works in your fly box. This is one of those patterns!

When designing representative fly patterns, there are some key features to consider. Do you intend it to float or sink, what size, colour, and footprint, along with any other outstanding characteristics, do you want? We try and tick all the necessary boxes.

This extended body is made from 1mm Dracon braided flyline backing which is marketed white or natural. I like to colour the Dracon backing, about a metre at a time, with waterproof felt pens. The best way to do this is as in my video. On a piece of white foam, lay down the Dacron and trap it under the blue felt pen by pushing down on it. Once the Dracon is trapped between the foam and the tip of the felt pen, pull the Dracon out between the foam and the pen. As it emerges, the colour will change from white to blue. You may have to do this a couple of times to achieve a good vivid colour.

Once you are happy with the blue base colour, you can draw on the black barring with a smaller black felt pen. Depending on what type of felt pens you have, drying time may vary before use. One metre with pre-coloured braided backing is enough for two dozen damsel flies. As you can see in the dressing image for this pattern, I also tie them in a brown olive colour to imitate the egg-laying female.

The hackle tips used for the wings of the adult damsel fly are best if they are a shade of dun, but off-white and even white will work, if you have these close to hand. Many stores and flytying shows have stalls that have cases of mixed colours of natural Indian cock and hen capes for sale. As these are generally reasonably priced in respect to other hackle products, I like to rummage in these cases as you can find some excellent capes in unusual natural colours for little money. Although not suited for dry fly hackle, I find these perfect for such jobs as wings.

When tying in the wings I find it prudent to keep the lower barbs on the hackle stems (*step 8*). This makes handling and attaching the wings much easier.

Regarding the parachute hackle, this should be a blue dyed grizzle cock hackle. I prefer to use a cock hackle here rather than a saddle, as there is more variation in barb length on a cock hackle, especially if you make use of the lower barbs. I also recommend a hackle that is considerably larger than standard for the hook size being used. This large hackle, extended body, wings and Ice dubbing all contribute to the overall buoyancy of the pattern.

The Ice dubbing should be the fine kind and spun in a dubbing loop made from split tying thread, which will give it optimal spikiness. It should also be left relatively long as this gives a larger surface contact and the impression of legs.

The foam for the post and thorax is cut in 3mm strips from a 3mm sheet. You may find it challenging sourcing the ideal coloured foam for the blue damsel, but don't worry, this is not critical and is more for the fisher than the fish. When you have a sheet, even a small sheet, once it's divided into 3mm strips, it's enough for several dozen flies.

TECHNIQUES MASTERED

Extended body
- Making an extended adult damsel body from braided Dracon backing line, for both male and female adult damsel flies.

All-in-one foam post and thorax
- How to use a simple foam strip for both the parachute hackle post, thorax and head of the fly.

Ice dubbing thorax
- Using a split thread dubbing loop technique which is useful for making spiky dubbing with lighter dubbings and materials.

Tying the Braided Blue Damsel

THE DRESSING

Hook: Mustad Heritage R30AP # 12

Tying thread: Sheer 14/0 black

Extended body: 1mm Dracon braided backing

Wings: Two dun cock hackle tips

Post / Thorax: Blue foam strip

Thorax/legs: Ice dubbing blue

WATCH THE VIDEO

youtube.com/watch?v=LUD7osSrw50&t=285s

Tying the Braided Blue Damsel with Barry Ord Clarke

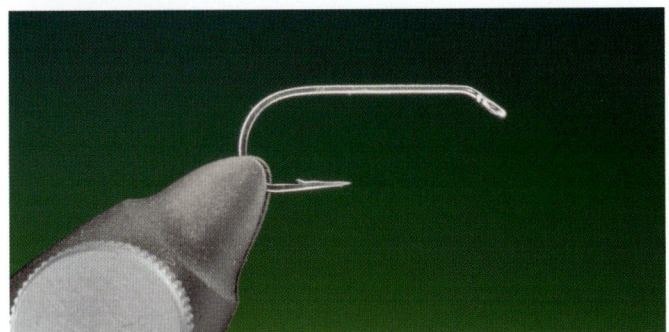

1 Secure your hook in the vice. Make sure that the hook shaft is horizontal. If you have a true rotary vice, centre the hook.

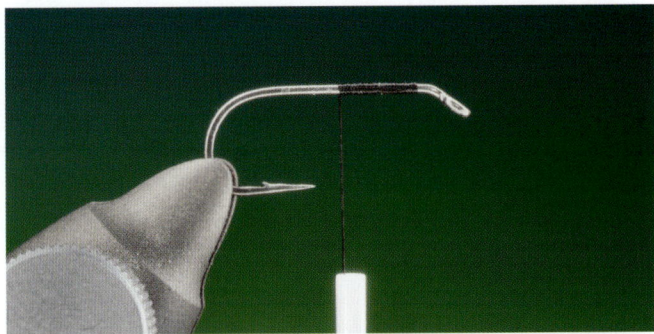

2 Attach your tying thread as shown and run a short foundation on the first half of the hook shank.

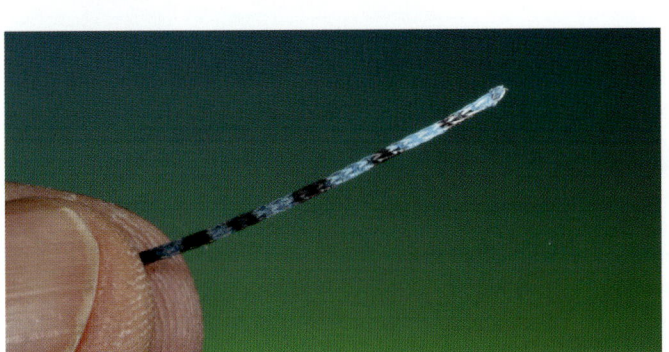

3 Cut an approximately 5cm length of the pre-prepared Dracon backing. Using a lighter or a cautery tool, carefully melt the very end of the braided Dracon. This will stop it from unravelling under casting.

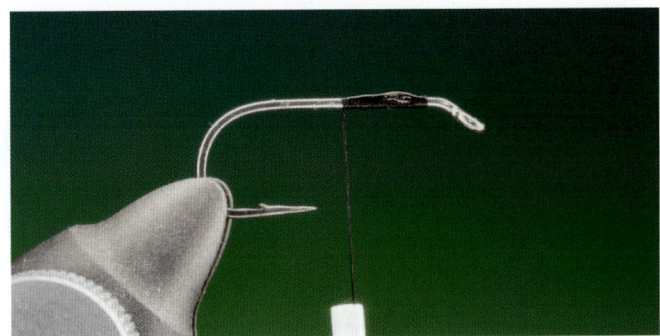

4 Place a small drop of superglue on top of the tying thread foundation. As the braid is quite slippery, this is just a precautionary step to ensure that the braid is firmly attached. I recommend using superglue gel rather than the less viscous glues.

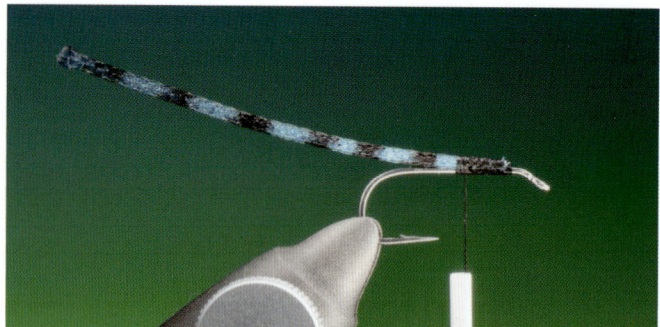

5 Tie in the extended body as shown.

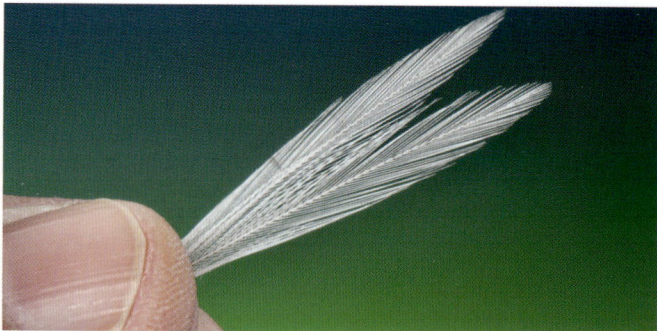

6 Select two similar sized dun cock hackles for the wings. The final wings should be just a little shorter than the body.

7 Position the first wing at the very rear of the black tying thread foundation and secure in place with a few wraps of tying thread. Adjust the length of the wing by gently pulling on the hackle stem.

8 Repeat the procedure for the second wing. Pay constant attention to proportions.

9 Once the wings are in position and secure, you can trim away the excess hackle stems and tidy up with a few wraps of tying thread.

10 From a sheet of blue foam, cut an approximately 3mm strip. Try and make the cut as even as possible; this will result in a more accurate finish.

11 Position the foam strip on top of the hook shank, over the thorax. Spin your bobbin holder anti-clockwise so your tying thread attains a flat profile. You can now secure the foam as shown.

12 Choose a nice blue dyed grizzle cock hackle, oversized for the hook size. If you don't have dyed grizzle, a blue dyed cock hackle will suffice.

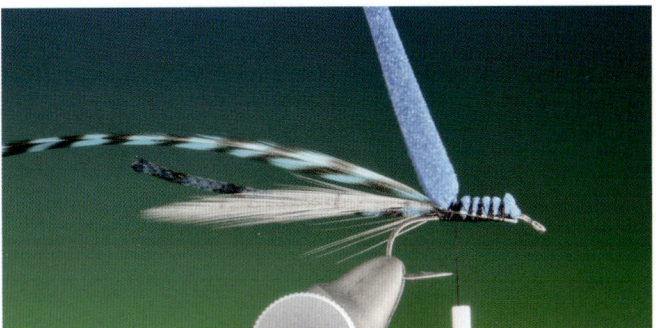

13 Tie the hackle in flat, at the very base of the foam post and secure the remaining hackle stem along the side of the thorax.

14 Attach a hackle plier to your hackle tip and wrap the whole hackle around the foam post to form a oversized parachute hackle.

15 Separate the barbs on the front half of the parachute hackle and tie off the hackle tip over the thorax. Trim away the excess hackle tip.

16 Spin your bobbin anti-clockwise so your tying thread attains a flat profile. Split the tying thread to form a dubbing loop and insert a little blue Ice dubbing.

17 Now spin your bobbin clockwise, so the loop spins a dubbing brush. Once the dubbing is spun, wrap the dubbing forward over the thorax towards the hook eye.

18 Secure behind the hook eye.

19 Spin your bobbin once again anti-clockwise so your thread has a flat profile and won't cut through the foam. Stretch the foam over the thorax, taking care that it's lying flat and in the correct position. Secure with a couple of wraps of tying thread.

20 Trim away the excess foam to form a little head over the hook eye.

21 Complete with a neat whip-finish or two, remove the tying thread and apply a tiny drop of head cement if desired. Trim away the longer Ice dubbing fibres on the underside of the thorax.

22 Bird's eye view of the finished adult Blue Damsel fly. Note the rearward fanned parachute hackle and long outward-reaching Ice dubbing fibres.

The Worm

Spinning fur and Ice dub into a dubbing brush • Soft swimming body • Heavy front shank head

There are many patterns known to saltwater seatrout and striper flyfishers to imitate sea worms, some better than others, some simple to tie and some not so simple. The original worm pattern 'Masken' from the tying bench of innovative Swedish flytyer Robert Lai is still, for me, one of the best, but it falls into the category of almost impossible to tie!

In more recent years there has been a popularity in tying articulated 'Game Changers' on 3, 4, 5 or even more shanks joined together to create a articulated swimming action. Although effective, these patterns are time-consuming to tie and heavy, not to mention expensive.

The pattern I tie here started off about 25 years ago as a simpler variant of Robert's original pattern, but over the years I have developed the technique somewhat and it is now easier to tie than ever before. But the basic original principle is still there. Few worm patterns swim and pulsate in the water quite like this one, imitating the natural swimming worm as closely as humanly possible with fur and steel. Very challenging to tie!

There are a few rules one must follow when tying this pattern. The tail hook should be light in weight. Because the worm has an extremely flexible body, a larger and heavier tail hook has a tendency to 'hang-up' on the body under casting, which results in you fishing a ball of fur with the business end out of line. A heavier tail hook also reduces the swimming action of the worm by restricting the tail from lifting when the bead head sinks.

The central core of the fly is the Dyneema loop that you spin the fur into and it is also the spine of the pattern which holds the front shank to the tail hook, so this loop is absolutely key to the success of tying this pattern. If the spine is not strong enough or securely attached to the front shank, you can risk losing not only the fish, but also the business end of your worm. So make sure that you tie this in as well as you can and don't be afraid to use superglue. If you would like a lighter and even more mobile worm, use marabou instead of fur. If you can't get racoon or opossum in the colours you would like you can use cross-cut rabbit but try and remove some of the underfur first; it will spin better.

The Latin name for the common ragworm is *Nereis diversicolor*, meaning they are quite variable in colour, but typically they are reddish brown, turning more on the green/blue side during the spawning season. So the rule for colour is that there is no rule: you can tie the worm in any colour you like! Personally I have found the two most successful colours for me are the one shown here and bright orange.

We are not 100% sure, and that's not for lack of theories, but the spring swarming (often and wrongly called a hatch) is due to the worm's spawning season and seems to be triggered by two main factors: a rise in water temperature of 6-7 degrees; and the arrival of a new lunar phase (full moon) anywhere around mid March and into April. The female ragworm broods her eggs within her long flattened body and as the eggs develop, her body becomes brittle and eventually splits, releasing the eggs. The male ragworms are attracted to the egg-laying area by following pheromones which are also released by the females. After spawning, both male and female ragworms die.

Ragworm swarming can be very localised, and it's not easy to know where. If in doubt then you should look to the sky, because the greedy and forever-hungry gulls can show you the way. If you can see that screaming seagulls are flocking and circling around an area of coast, this shows you where to fish – just like the pelicans indicated tarpon fishing. Consider also when the strong spring sun has been high in the sky all day and warming up the shallows, especially with dark muddy bottoms. Most seatrout flyfishers, including myself, prefer sight fishing during the day, looking for rises as you fish your way systematically through all possible holding spots in small bays and inlets as the tide rises and falls. And if you are, like most seatrout flyfishers, hoping to connect with larger fish that are normally wiser and more sceptical about entering the shallower coastal waters during the hours of daylight, these shallow areas retain the day's heat during the first couple of hours of darkness. It's during this period that larger seatrout dare to venture into the shallows to feed. You should then fish at least a couple of hours into darkness.

And don't forget that worms are not only on the saltwater menu. All fish eat worms all year round. This has proven a deadly pattern for me, for most predatory fish, in both fresh and saltwater!

TECHNIQUES MASTERED

Spinning fur and Ice dub into a dubbing brush
- How to mix fur and Ice dubbing and spin in a dubbing loop to form a flashy fur dubbing brush.

Soft swimming body
- Using only Dyneema or GSP tying thread to make a soft yet ultra strong animated body without articulation.

Heavy front shank head
- A hook-free head with heavy slotted Tungsten bead to provide the undulating swimming action.

Tying The Worm

THE DRESSING

Hook: Mustad Heritage C49 # 8-10

Tying thread: Dyneema 110 or equivalent GSP thread

Tail: Racoon zonker and Pearl Ice Dub

Head Shank: Fish Skull articulated shank 15-20mm

Head: 4.6mm slotted Tungsten bead

Body: Racoon zonker and Pearl Ice Dub

WATCH THE VIDEO

youtube.com/watch?v=b0XYkvdUsYE

Tying The Worm with Barry Ord Clarke

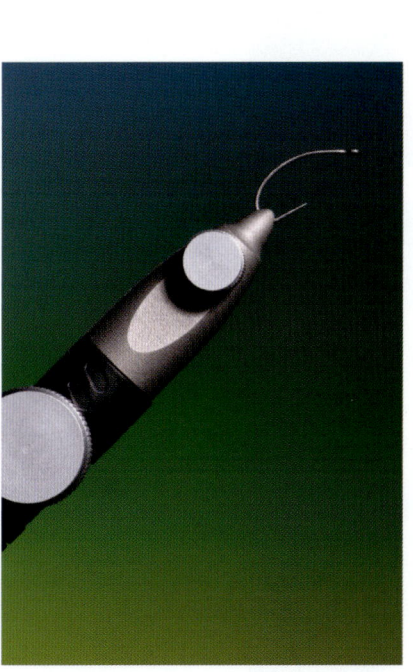

1 Secure your lightweight tail hook in the vice. If you are tying for salt or fresh water choose your hook accordingly.

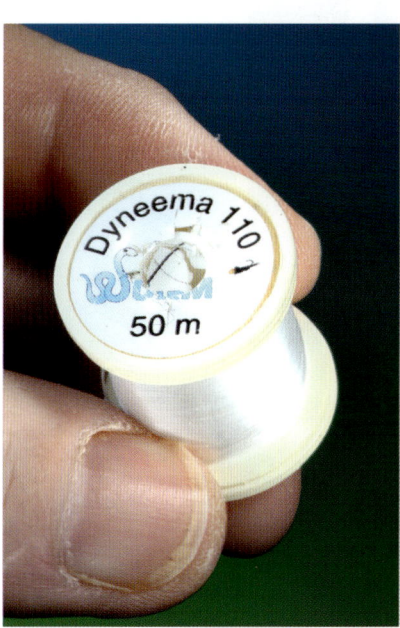

2 You must have Dyneema 110 or a GSP thread of an equivalent strength.

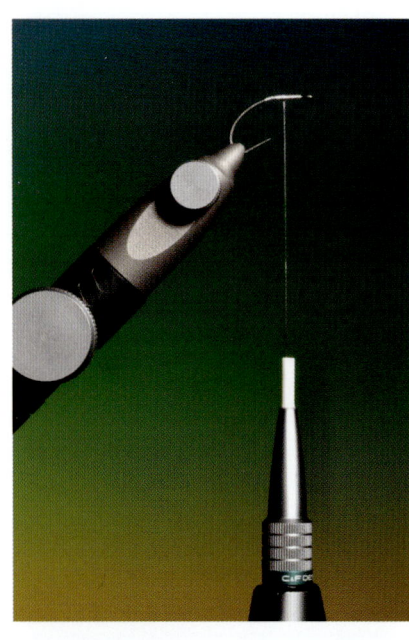

3
Cover the front half of the hook shank with a foundation of Dyneema tying thread.

4
You will now need a heavy dubbing spinner. Don't try and use a wishbone dubbing spinner as it won't work with thick fur!

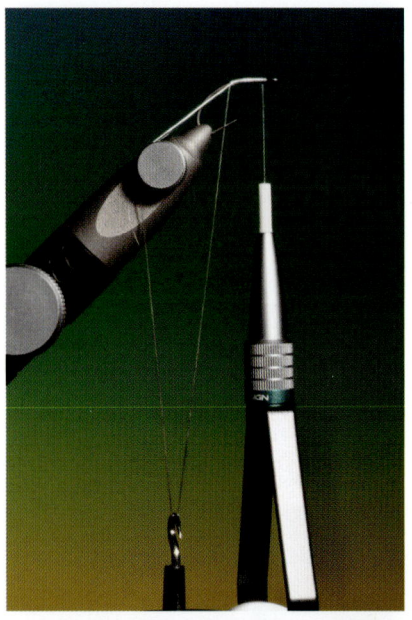

5
Make a short dubbing loop at the rear of the tail hook and run your tying thread forward again to the hook eye. Hang your dubbing spinner in the loop.

6
Cut a short length of fur strip, enough to fill a medium Petitjean magic clip. If you have it, a crosscut zonker strip is the easiest to use for this.

7
The Ice dub that you use should be fine: heavier Ice dub will not spin well.

8
Holding the strip as shown, pull just a little Ice dub from the packet to cover the strip. See video for this technique!

9
Now place the fur and Ice dub into the medium Petitjean clip, leaving enough room for the cut between the hide strip and the clip jaws.

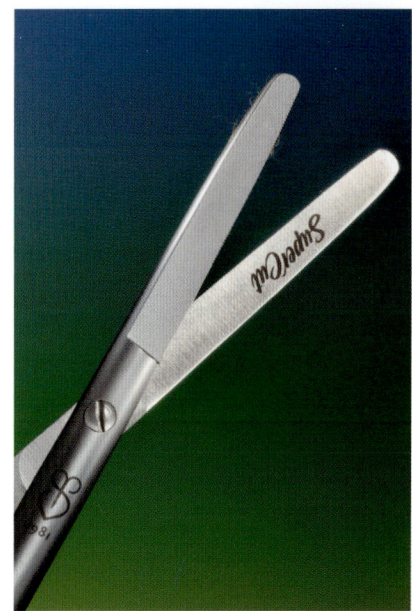

10
When cutting the hide strip away, I recommend that you use long straight scissors, so this is done with one simple cut rather than several short cuts.

11
Cut the hide away from the fur, resulting in a straight, even cut as shown.

12
Holding the dubbing loop open with your left hand, run a little wax on one side of the loop then place the fur in the loop. Now the hard part! Once the fur is trapped in the dubbing loop, don't release the tension.

13
Spin your dubbing spinner clockwise to form a dubbing brush. Once the brush is tight, use a brush to free up any trapped fibres.

14
Now wind on the dubbing brush so as to form the tail of the worm. Make sure that with each wrap of dubbing, you brush the fibres backwards. Cover the whole hook shank.

15

Whip-finish and remove your tying thread. Brush out the fibres again. Give the head a coat or two with varnish and remove from your vice.

16

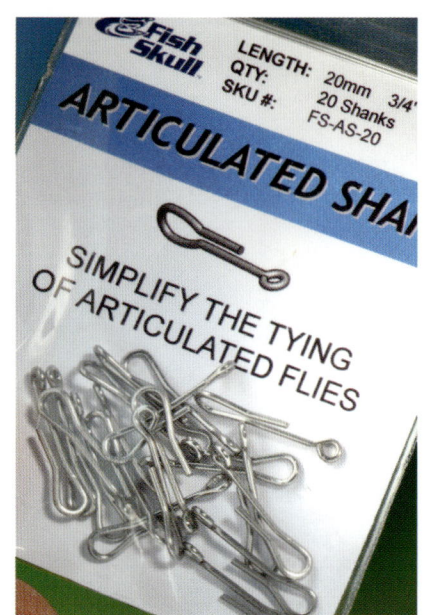

For the front of the fly you will need a 15 or 20mm shank. Some tyers use a regular hook for this, but I find this only causes problems when casting and fishing, increasing hang-ups and snags.

17

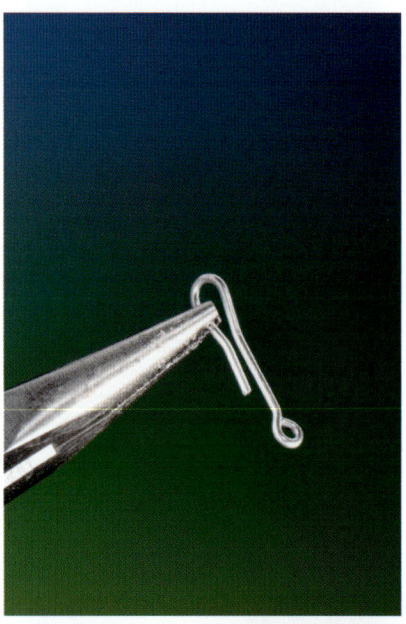

Hold the shank as shown with pliers.

18

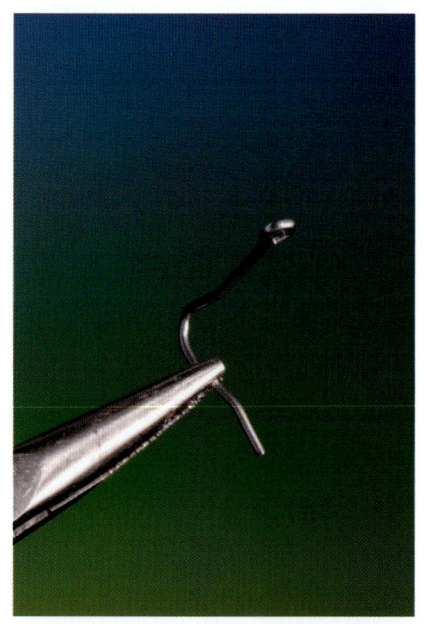

Open the shank up by bending it just enough to get the bead on. Try and keep the whole shank 'in line' when opening and closing.

19

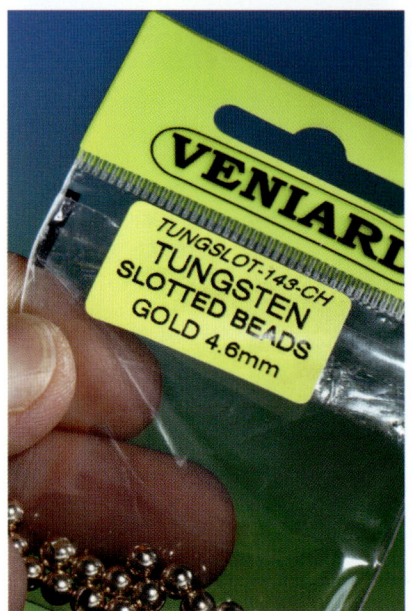

You will need a heavy slotted Tungsten bead for the head. The heavier the bead, the more animated the swimming action.

20

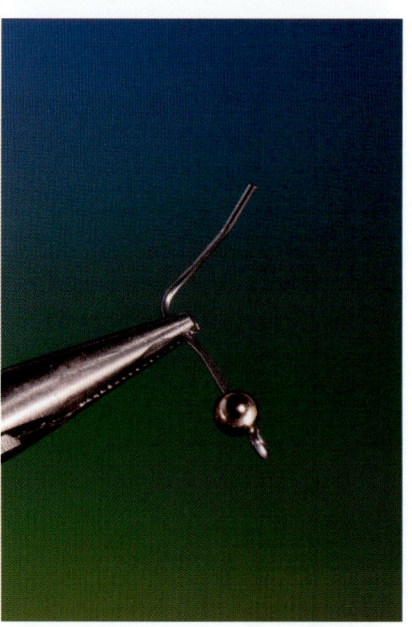

Place the bead on the shank, slot to the rear.

21

Close the shank as much as you can before securing it in the vice.

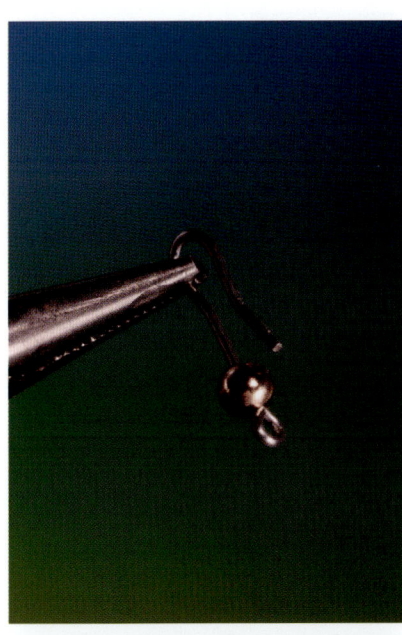

22

Secure the shank in the vice as shown and attach your tying thread. Once you have made a few wraps around the shank you can slowly start to tighten each new wrap to close the shank. Secure forward to the bead head.

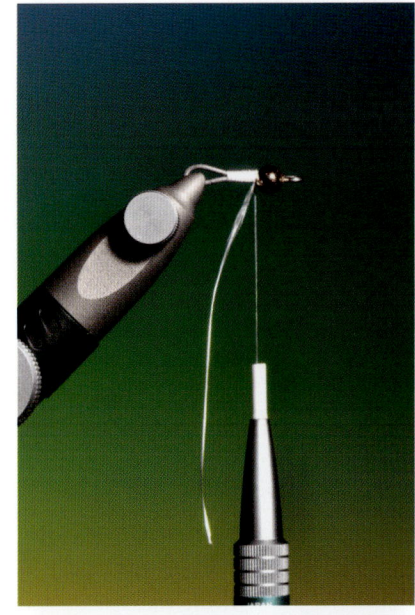

23

Trim away the surplus tying thread.

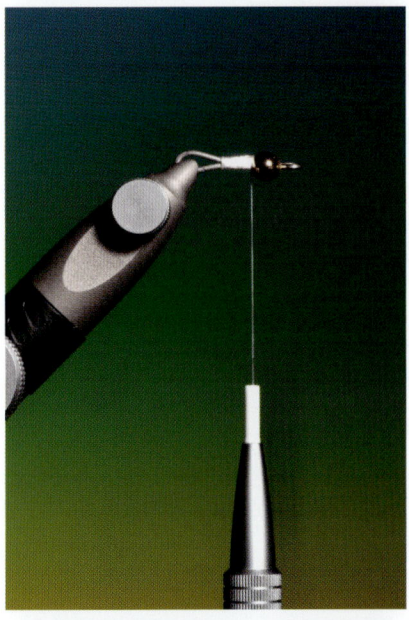

24

Give the whole shank a coat with cement or superglue.

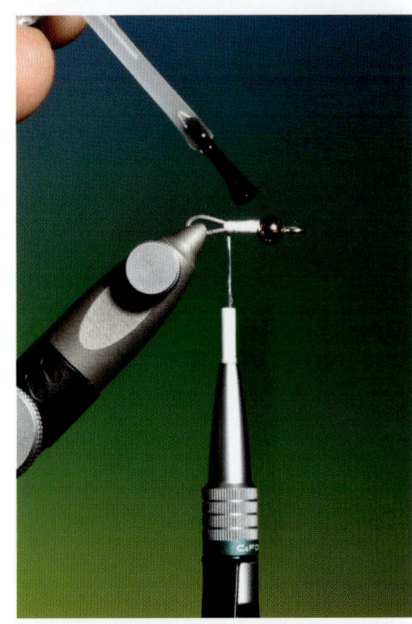

25

Make a whip-finish at the rear of the shank as shown.

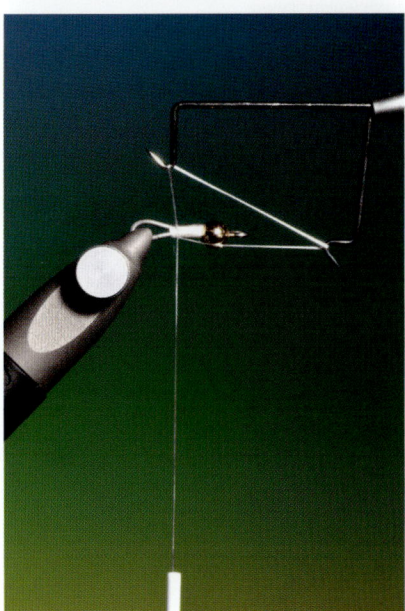

26

Now make a large dubbing loop at the rear of the shank and finish with your tying thread by placing one wrap between the bead and shank eye. This you will wind back later.

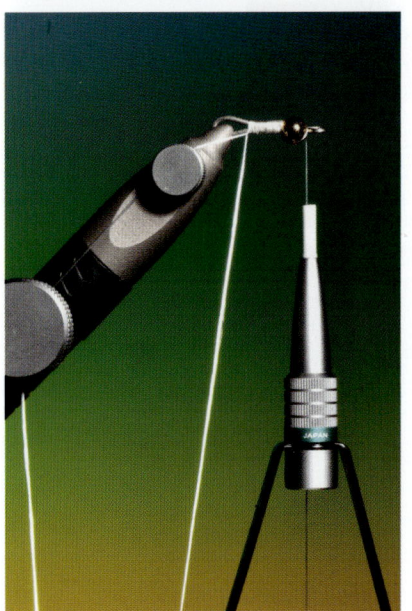

27

Using a floss threader: see video. Thread your tail hook onto the floss threader as shown.

28

Now thread the dubbing loop through the loop on the floss threader.

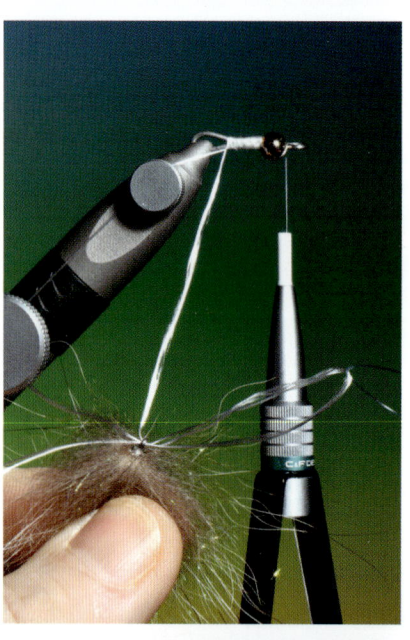

29

While holding the tail hook in your left hand, pull the floss threader with your right hand so as to pull the dubbing loop through the tail hook eye.

30

The tail hook should now be on the dubbing loop.

31

Load two larger Petitjean magic clips with fur and Ice dub, as you did with the first.

32

Here is the tricky part. Wax one side of your dubbing loop. Place the fur from one clip in the top half of the loop.

33
Keeping the tension on the loop, slide the tail hook up to the first fur strip. This will keep the top half of the loop closed. Then open the bottom half of the loop and place the second material clip below the tail hook.

34
While holding the loop closed at the tail hook point, spin the bottom half of the loop. Once this has spun, release your grip on the tail hook and the upper half of the loop will spin automatically. You can now spin the whole loop a little more, to tighten securely.

35
While holding the loop out-stretched and tight, use a toothbrush (not a metal dubbing brush! This will fray and weaken your Dyneema) to open out any trapped fibres.

36
Take hold of the tail hook with your left hand and the dubbing spinner in the right, keeping everything tight, then bring your tying thread back to the rear of the shank. Make a couple of turns of tying thread to hold it in place, and remove your dubbing spinner.

37
Now release the tail hook from your left hand and the two cores of the dubbing loop will naturally spin around each other and form a secure and strong spine to the fly. If it only spins a little, you can help it by spinning it by hand, but only in the same direction as the natural spin! Give it a comb to make it spin more.

38
Secure the loop on the shank by folding it forward and back as you go to secure it correctly. Give the whole shank a coat of superglue.

39 Now make a small dubbing loop at the rear of the shank.

40 Load with fur and Ice dub as you did for the tail hook.

41 Spin up the dubbing loop into a dubbing brush.

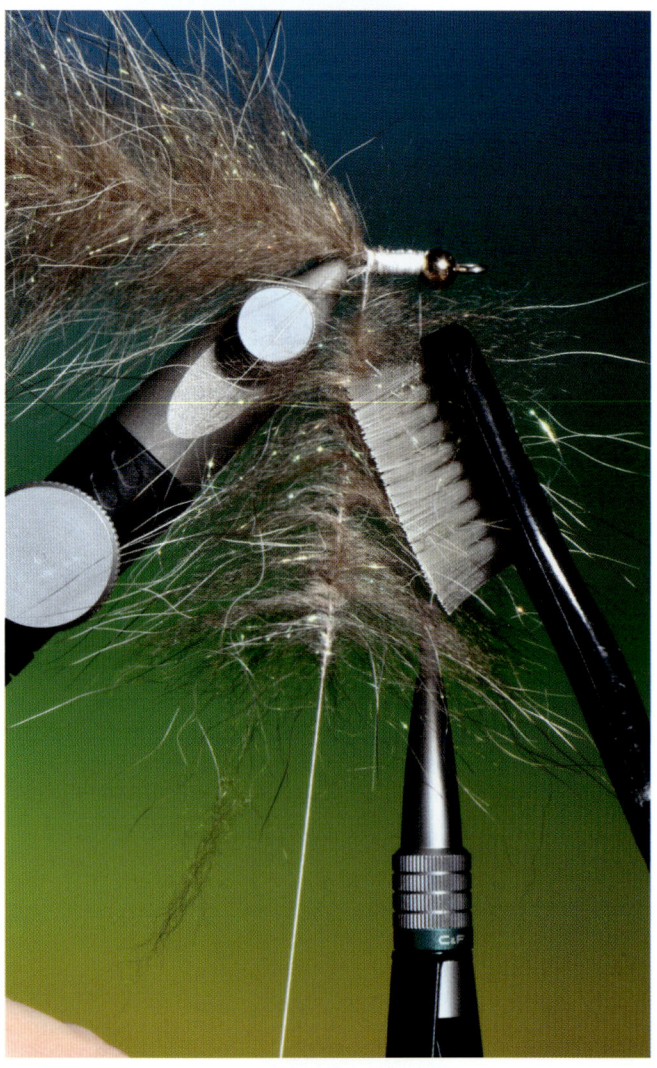

42 Brush out any trapped fibres as before.

43

Wrap the dubbing brush over the shank, as you would a wet fly hackle, covering the whole shank, and remove the dubbing spinner. Tie off behind the bead head.

44

Whip-finish and remove your tying thread. You can now varnish the head by placing drops of varnish directly onto the shank eye in front of the bead as shown. You will see the varnish drain into the bead head. Repeat a couple of times. This will secure the the bead and the body to the shank. Clean the eye of the shank by passing a surplus hackle through the eye.

45
The finished version of The Worm which will take fish anywhere!

㉓

Float Foam Ant

Foam body • Hackle tip wings • Deer hair leg construction

From the family *Formicidae*, ants are without doubt the most important terrestrial with regards to the flyfisher, with nearly 14,000 ant species which have colonised almost every single landmass on earth. The few places without indigenous ants are Antartica, Greenland, Iceland and a few other inhospitable environments. They range in size from 0.75-55mm and vary in colour from brown-black to yellow-red.

Since its introduction in the early 1980s, the use of foam within flytying has been both loved and abhorred. The state of the art flytyer embraces both its versatility and its ability to add buoyancy to not only the smallest dry flies, but also the largest saltwater predator patterns. On the other hand, the traditionalists amongst us look upon its use as almost heresy: an abomination! Either which way you lean, foam is most definitely here to stay.

Naturally one of the earliest insects to be imitated with foam was the ant. To date there are too many foam ant patterns to name, varying from the ultra-realistic to the abstract (the Chernobyl ant). The

Float Foam Ant is one of my favourites, which as far as I know, is a variant of Rainy's original ant.

Float foam from Rainy's Flies in the US first started the foam revolution in flytying. Its cylindrical form and range of sizes makes it a great choice for ant and beetle bodies. Although there are many contemporary patterns that are constructed of foam only, I am personally of the school of mixed media, a fusion of both natural and synthetic, with both playing their role in fly design.

Please take note that when tying in any foam, use thread control to avoid cutting the foam when applying pressure with your tying thread. This is done by letting your bobbin holder hang and then giving it a spin, anti-clockwise. This will give your tying thread a flat profile. The advantage of this is a broadened (flat) pressure when applied to the foam. A round profile on the other hand will result in a focused pressure which will cut into or through the foam.

If Rainy's Float Foam is not available, you can use regular foam cylinders or even foam strips cut from a sheet. When using foam, choose the correct foam for the correct job at hand. Some foams are sponge-like and absorb water, and these should be avoided for obvious reasons. Look at the foam density, rigidity and flexibility, all important factors to consider when designing flies.

You can also substitute the ostrich herl with peacock herl, if you would like a more iridescent metallic finish to the abdomen.

> **TECHNIQUES MASTERED**
>
> **Foam body**
> - How to use and control your tying thread for correct use with foam. This will ensure that your thread will not cut or damage the foam when tied in.
>
> **Hackle tip wings**
> - How to prepare and correctly mount two hackle tip wings in the correct position for a flying ant.
>
> **Deer hair legs**
> - Selection and preparation of deer hair technique for using a small bunch of deer hair to create realistic buoyant insect legs.

This is a cylindrical extruded foam with a slightly silk matt finish. It's available in four sizes: small, medium, large and X-large and in several colours.

Tying the Float Foam Ant

THE DRESSING

Hook: Mustad Heritage C49S # 10-16

Tying thread: Sheer 14/0 black

Body: Black float foam and black ostrich herl

Wings: Two Chocolate dun cock hackle tips

Head: Black float foam

Legs: Deer hair dyed black

WATCH THE VIDEO

youtube.com/watch?v=54nOc62CNA8&t=111s

Tying the Float Foam Ant with Barry Ord Clarke

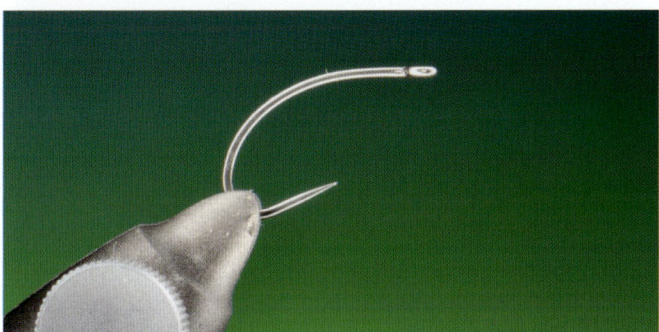

1 Secure your hook in the vice. Make sure that the hook shaft is horizontal. If you have a true rotary vice, centre the hook.

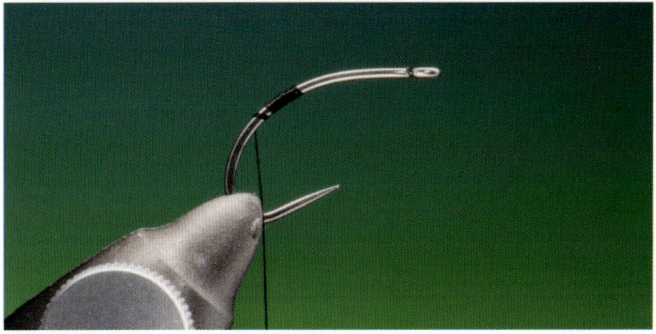

2 Attach your tying thread as shown and run a short foundation down into the hook bend.

3 Cut a short length of small black float foam.

4 Using long straight scissors, make one single cut down the centre of the foam piece, cutting it into two.

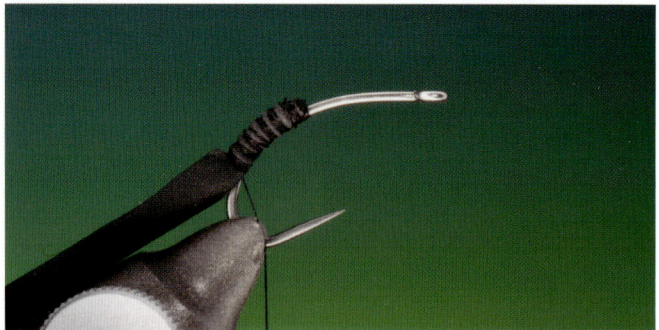

5 Using only one half of the foam cylinder, attach this, cut side up to the rear of the hook shank. Take special care to the positioning of the foam on the hook shank!

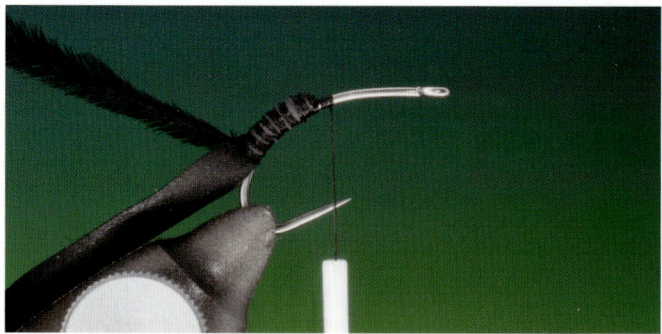

6 Select a nice dense black ostrich herl and tie this in with the heel facing backwards as shown.

7 Now attach hackle pliers to the herl and wrap forward in tight even turns, taking care not to twist the herl.

8 You can now reverse-wind your tying thread and tie off the herl. Cut away the excess herl.

9 Comb the herl slightly back and tie down with a few turns of tying thread until you have the correct abdomen size.

10 Spin your bobbin anti-clockwise, so it has a flat profile. This will stop the thread cutting through the foam when tying it down. Stretch the foam over the ostrich herl and secure with three or four tight wraps of tying thread.

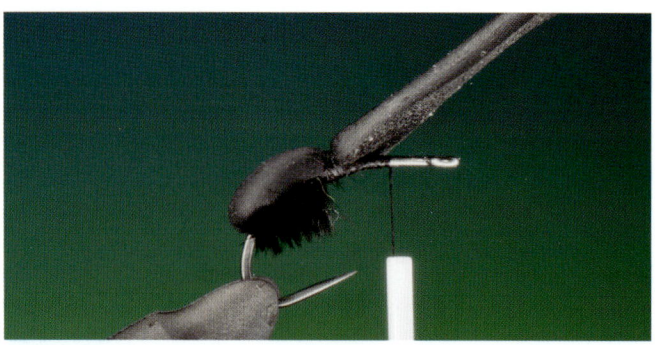

11 Once the foam is secure, wrap your tying thread forward under the foam as shown.

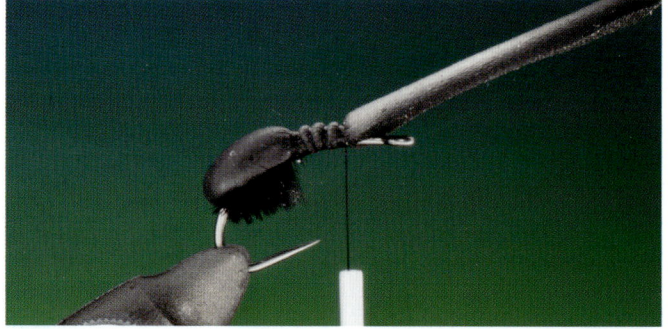

12 Secure the foam over the centre of the hook shank.

13 Trim away the excess foam.

14 Take the remaining foam half strip and tie in as the first. Once again take note of the position on the hook.

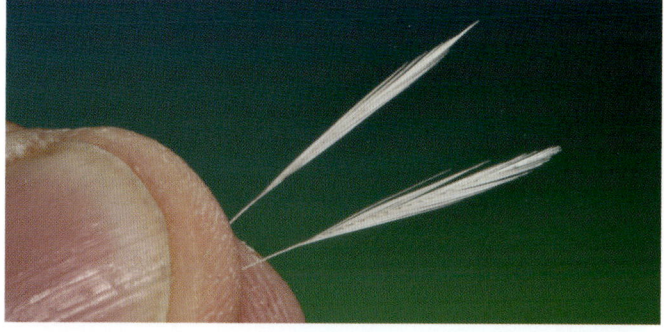

15 Select two chocolate dun cock hackles. Measure the hackle tips to the correct length and strip away the lower barbs, to form the two wings.

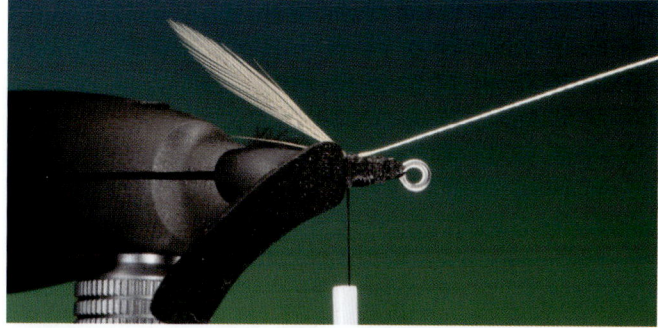

16 Tie in the first wing tight into the forward foam strip. It's advisable to apply a little wax to your tying thread here: this will stop the hackle stem from slipping.

17 Now tie in the second wing, so that it mirrors the first. You can adjust the length of the wings by carefully pulling on the hackle stems.

18 Once the wings are correctly positioned, trim away the hackle stems.

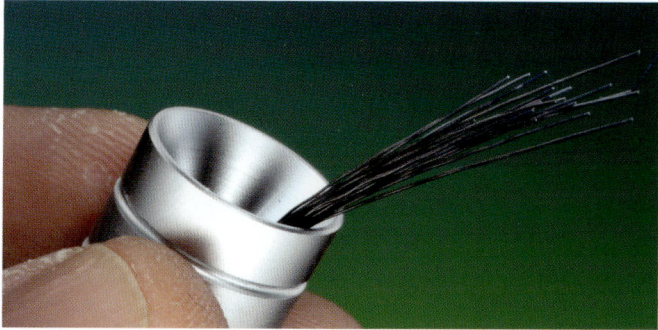

19 Cut a small bunch of black dyed deer hair. Comb out all the underfur and short hairs before stacking. You'll only need about 10-15 hairs.

20 With the stacked tips facing backwards, tie in the deer hair over the foam strip, approximately the same length as the body.

21 Once secure you can trim away the surplus deer hair and tie down the ends.

22 Using a dubbing needle, separate the deer hair, half and half each side of the thorax, 90 degrees from the hook shank. Fold the foam over the thorax so it holds the deer hair in position. Secure with a few turns of tying thread.

23 Using fine sharp scissors, cut away the surplus foam behind the hook eye.

24 Starting at the hook eye and working your way back, build a neat head with your tying thread. Whip-finish and remove your tying thread.

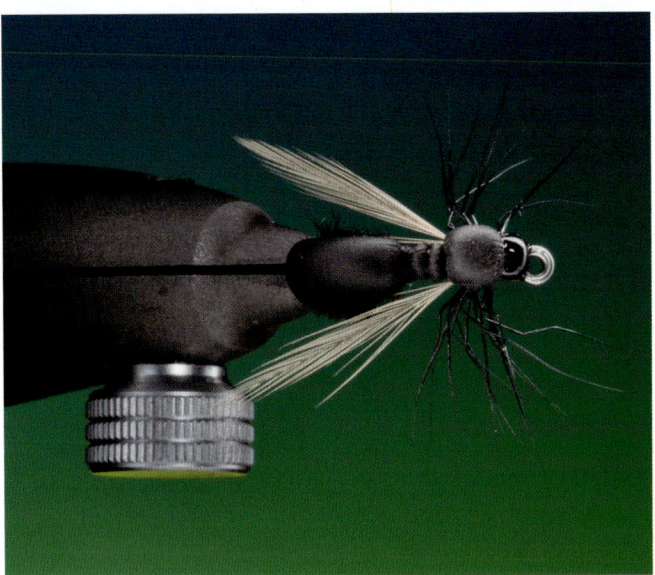

25 Watch the float foam ant video for the leg technique. Give the head of the ant a coat or two of varnish.

26 The finished Float Foam Ant.

24

Madam X

Floss silk body • Deer hair bullet head and wing • Double rubber legs

If you have never had a relationship with Madam X, it's about time you got to know her.

The Doug Swisher pattern, Madam X, falls under the category of attractor and searching pattern, but it's a fly that can work under many different circumstances and hatches.

The amount of deer hair used in the underbody, tail and wing makes Madam X an extremely buoyant pattern, and the four very visible and mobile white rubber legs certainly make enough disturbance in the surface to get the attention of even the most lethargic trout! As with the original pattern, I like to use bleached deer hair. This is especially helpful for failing eyes, after sundown. Proportions are particularly important for getting this pattern balanced with the correct wing length and head size.

This can easily be achieved by firstly measuring the wing from directly behind the hook eye to the end of the tail. Once the hair bunch is cleaned and stacked, hold the hair in your left hand and, using your thumb as a marker, measure from the tip of the tail to the hook eye. Move the bunch forward and tie in by your thumb behind the hook eye.

Secure the bunch tight into the hook eye, making sure that the hair for the bullet head has spun around the whole hook shank. You can then wrap your tying thread back over the butt ends of the deer hair, finishing exactly where the bullet head will finish. This will be your marker, once the surplus deer hair is trimmed off. But remember when tying down the head to spin your bobbin anti-clockwise, so your tying thread has a flat profile, otherwise the tying thread will cut through the deer hair when you apply pressure.

Traditionally the wing is now folded over and tied down to form the wing and head, but if you use a small plastic tube to push back the wing as illustrated in step 16, you will get a perfect wing and tight bullet head. I prefer to use a transparent plastic tube; this allows you to observe what is actually going on with the deer hair. Is the hair even and straight? If not, just twist the tube a little to position the hair or remove it altogether and start again until it is right. I use clear plastic tubes for many flytying techniques.

The rubber legs are the attractor factor in this pattern. Don't make the legs too long, or they may catch around the hook bend when casting and disable the swimming effect. Although white is the original, many tyers opt for hi-vis coloured legs and even striped rubber legs. I have found that after some time, the rubber legs tend to get dry and a little stiff and on occasion have a shorter life than the deer hair and hook. If this is the case, and you are penny-wise, you can easily remove the old weathered rubber legs and replace them with new ones.

TECHNIQUES MASTERED

Floss silk body
- Using floss silk for a smooth full body of any colour.

Deer hair bullet head and wing
- A nice technique for making perfect bullet head flies from a small bunch of deer hair.

Double rubber legs
- A simple and effective method for tying in and cutting rubber legs to the correct length.

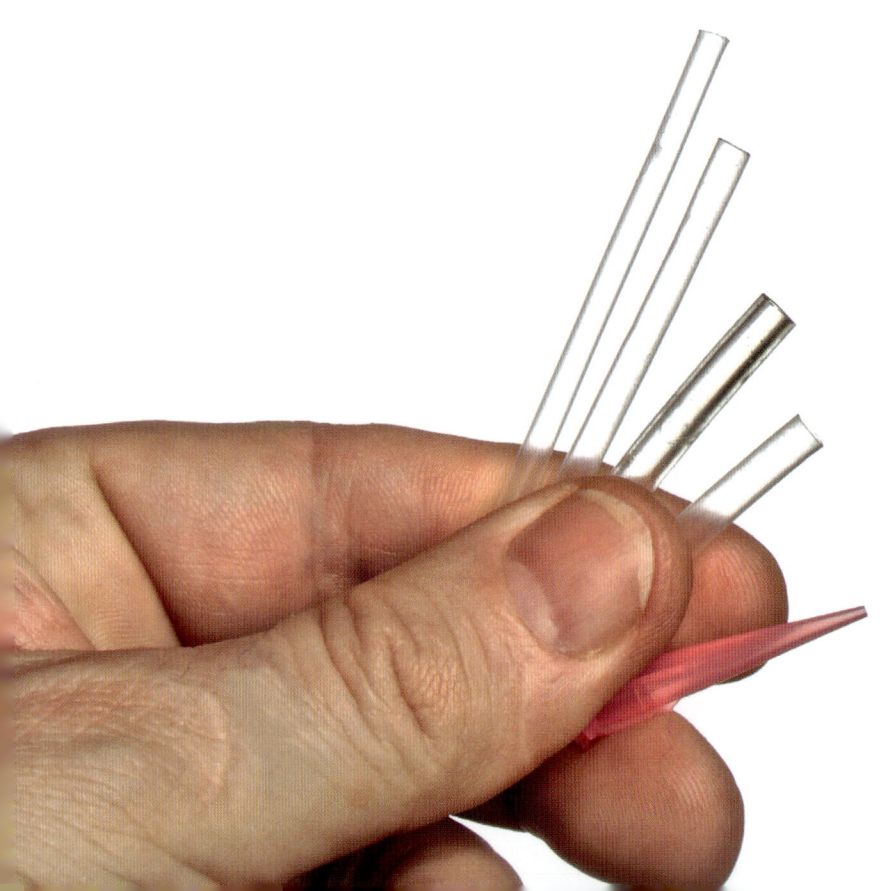

Tying the Madam X

THE DRESSING

Hook: Mustad Heritage R43 # 12

Tying thread: Dyneema 55 or equivalent GSP

Tail: Bleached deer hair

Body: Silk floss, the colour choice is yours.

Wing/head: Bleached deer hair

WATCH THE VIDEO

youtube.com/watch?v=k8VbjjhzIAc&t=62s

Tying the Madam X with Barry Ord Clarke

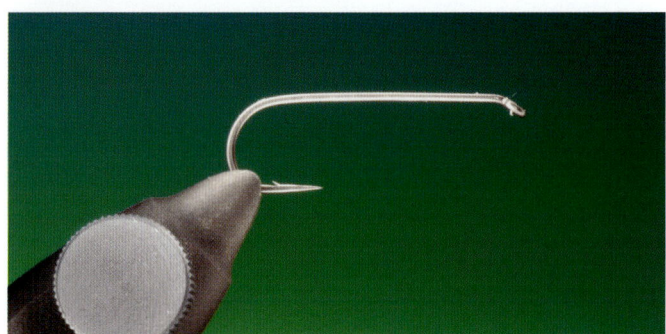

1 Secure your 3 extra-long dry fly hook in the vice. Make sure that the hook shaft is horizontal. If you have a true rotary vice, centre the hook.

2 You will need some Dyneema or GSP tying thread.

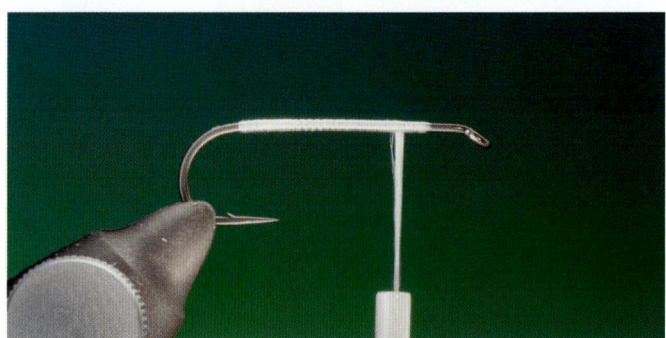

3 Attach your tying thread and run a foundation over the whole hook shaft as shown.

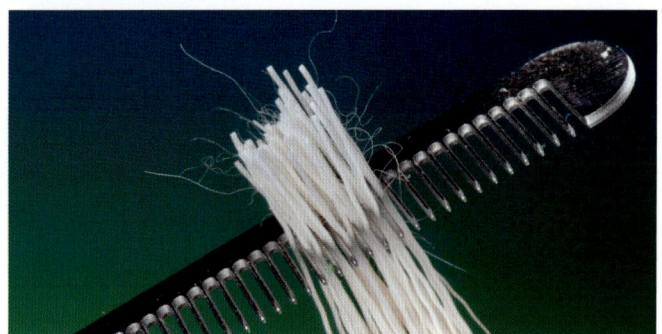

4 Cut a small bunch of bleached deer hair with fine unbroken tips. Clean the hair by removing all the underfur and shorter hairs.

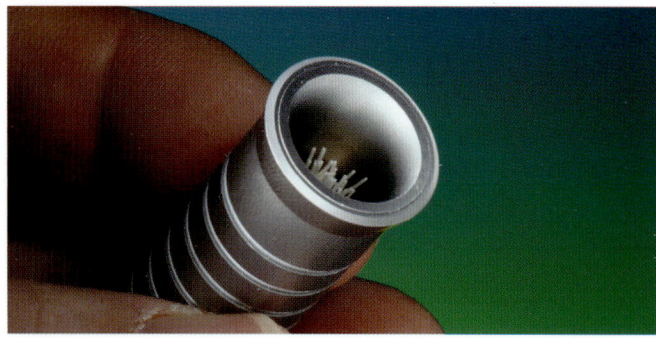

5 Place the cleaned bunch in a hair stacker and even up the tips.

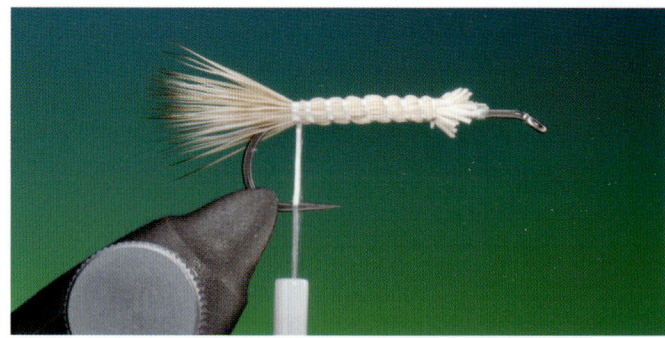

6 Remove the hair from the stacker. While holding the hair, measure the tail length and tie in as shown.

7 Spin your bobbin anti-clockwise so your thread attains a flat profile, then build up a even body with tying thread.

8 You will need some silk floss for the overbody. Here I have chosen metallic copper.

9 Attach your silk floss securely, at the front end of the body.

10 Keeping the silk floss flat at all times, wrap a relatively flat, even body, once up to the tail base and then back again. Tie off the floss silk.

11 Now cut, clean and stack a slightly larger bunch of bleached deer hair, so the tips are even. This is important if you wish to have a symmetrical wing.

12 Measure the wing along the hook shank. It should be as long or a little longer than the whole hook and tail length together. Tie this in tight into the hook eye. Making sure that the hair spins around the whole hook shank.

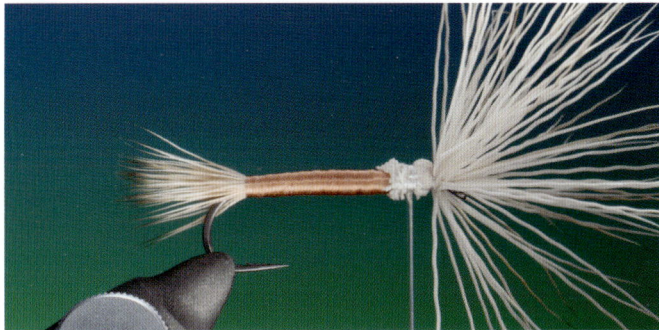

13 Secure tightly with wraps of tying thread, and then trim away the excess hair as neatly as possible.

14 Tie down the ends. This finished section should NOT exceed the size of the finished bullet head.

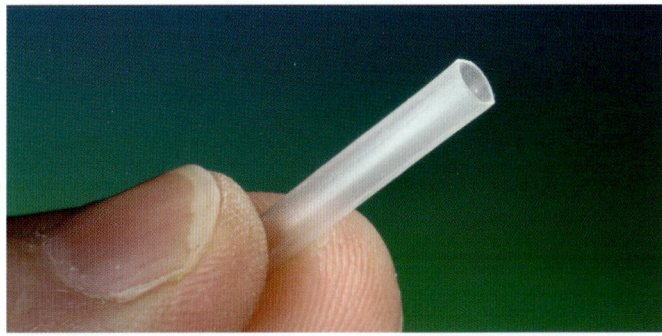

15 You will now need a small plastic tube. I like to use a transparent one, so I can observe what is going on with the hair through the tube.

16 Place the plastic tube over the hook eye and push all the deer hair back over the body. Make sure that all the deer hairs are lying correctly.

17 Keeping the tube on the hook, rotate your vice and check that the deer hair on the underside does not cover the floss body.

18 If you are happy with the deer hair in the head and wing, just where the tube ends, make a few hard wraps of tying thread to form the head.

19 Once secure, remove the tube and colour your tying thread with a red waterproof felt pen.

20 You will need some white round rubber legs.

21 Tie in the first legs on the side of the red collar. Take care that the legs are long enough. They can always be shortened, but not made longer.

22 Repeat with the second rubber legs. Make sure that they are balanced on each side.

23 Once the legs are secure you can make a couple of whip-finishes and remove your tying thread.

24 Trim the legs down to the desired length. The rear legs should be a little longer than the front legs.

25 View from the underside of Madam X.

26 The finished Madam X

27 Here is one tied with natural deer hair. It works just as well but is a little more difficult to see at distance on the water.

Foam Cylinder Crane Fly

Foam extended bodies • Knotted pheasant tail legs • Hackle tip wings

Firstly, let's get something out of the way, so as to avoid any confusion. In North America, daddy-long-legs is the name given to the wingless, harvest master spider, *Pholcidae*. But in Europe, daddy-long-legs is the name given to members of the crane fly family. It's all a little confusing but now that we are all on the same page…

Tipulidae, crane flies or daddy-long-legs, are a familiar sight both on and off the water more or less from early spring to late summer. There are in fact several hundred species of crane flies ranging in size from just a couple of millimetres to over 60mm long. Although most species of daddy are terrestrial, there are a few that are aquatic. Crane flies with their scraggly demeanour are remarkably poor aviators and once airborne are largely at the mercy of the wind and where it takes them, being forced to crash land on the water. With a decent breeze they can be blown across the water, like tumbleweed, trailing their legs behind them, in some cases even making a bow wave as they blow and skate across the surface. This behaviour and their sheer size make them extremely attractive to feeding trout and can provide the dry fly fisher with some memorable sport with violent takes.

Many detached body crane fly patterns are somewhat delicate and easily damaged, be it by fish, prolonged casting or just general wear and tear. Here is a pattern that has a subtle fusion of both natural and synthetic materials, a combination I have grown extremely fond of in my senior years.

Be very particular when choosing the foam that you are going to use for the extended body. Not all foam will float and some foams even absorb water! When it comes to knotting the pheasant tail fibres for the legs, I recommend that you avoid using short pheasant tail fibres: the longer the better. I also suggest that you watch the video via the link. Although an easy enough technique, it's a technique that is best demonstrated via video.

A good rule of thumb when purchasing pheasant tails from your dealer is to look through them *all*! They are normally displayed in store as individual packets or displayed in a vase like a bouquet of flowers. These are, as a rule both cock and hen pheasant tail feathers mixed together. What you want to look for is the 'centre cock pheasant tail feathers' and there are only one or two found on each cock bird. These are the longest, broadest and have super long fibres. The hen pheasant tails are a much lighter brown and shorter in length than the cock pheasant tails.

TECHNIQUES MASTERED

Foam extended bodies
- Choice and preparation of foam cylinders for extended detached crane fly bodies.

Knotted pheasant tail legs
- How to knot singular cock pheasant tail fibres to make realistic jointed legs.

Hackle tip wings
- Dry fly winging technique for attaching and rising of hackle tips for large double wings.

Tying the Foam Cylinder Crane Fly

THE DRESSING

Hook: Mustad Heritage C49XSAP # 14

Tying thread: Sheer 14/0 olive

Extended body: 2.3mm foam cylinder

Legs: Six or more knotted cock pheasant tail fibres

Body hackle: Natural brown saddle hackle

Wings: Two fine brown Indian hackle tips

WATCH THE VIDEO

youtube.com/watch?v=xLHZ83nK1rg

 Tying the Foam Cylinder Crane Fly with Barry Ord Clarke

 rb.gy/gd5lae
How to tie Daddy & Hopper Legs with Barry Ord Clarke

1 In this sequence I am using 2.3mm olive foam cylinders, but the choice of size and colour is entirely up to you. I know several flyfishers that swear by hi-vis colours, especially red and orange.

2 These Veniard foam cylinders are the perfect length for making two bodies from each cylinder.

3 While holding the foam cylinder in one hand and a lighter in the other, carefully warm the end, taking care not to burn it! While still warm, roll the end tightly between finger and thumb. Do both ends.

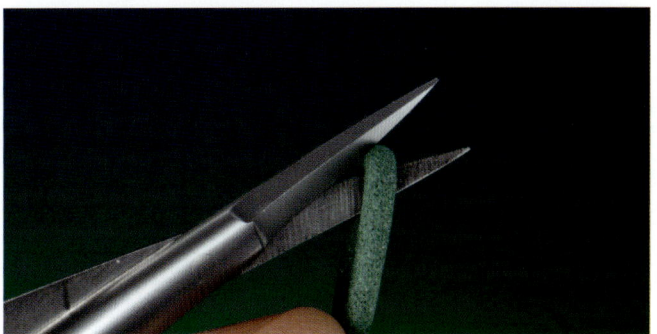

4 Fold the foam cylinder in half, end to end, and cut at the mid point to make two extended bodies.

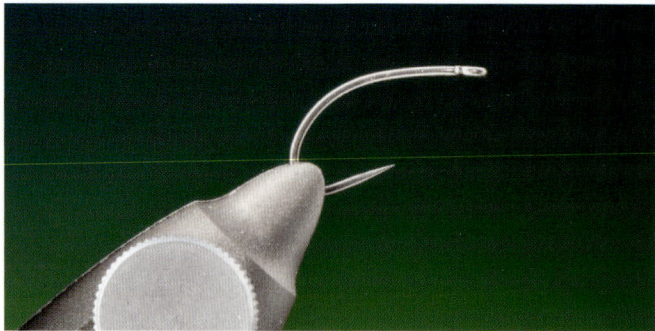

5 Secure your hook in the vice. Make sure that the hook shaft is horizontal. If you have a true rotary vice, centre the hook.

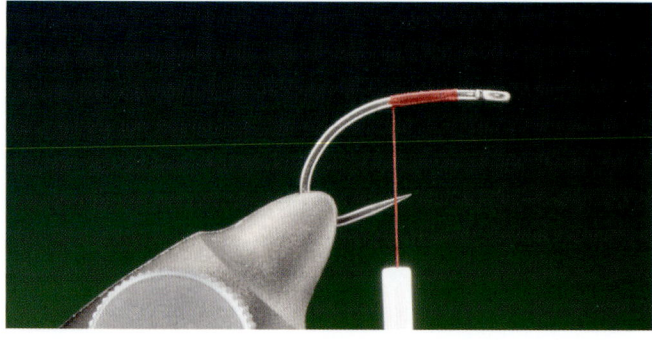

6 Attach your tying thread as shown, a little behind the hook eye and run a foundation a short way along the hook shank. The length of this foundation will be your guide for proportions.

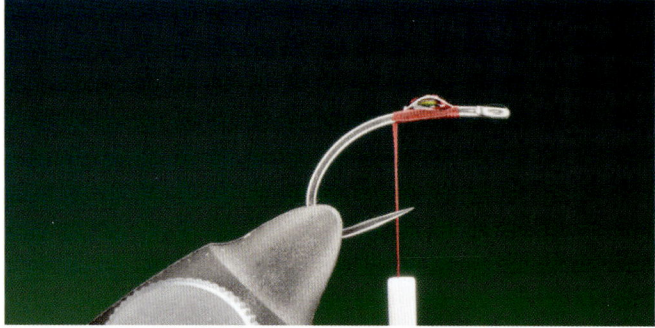

7 Place a tiny drop of superglue up on the short tying thread foundation. I prefer to use the superglue gel for tying as it's more precise on application and dries clear and flexible.

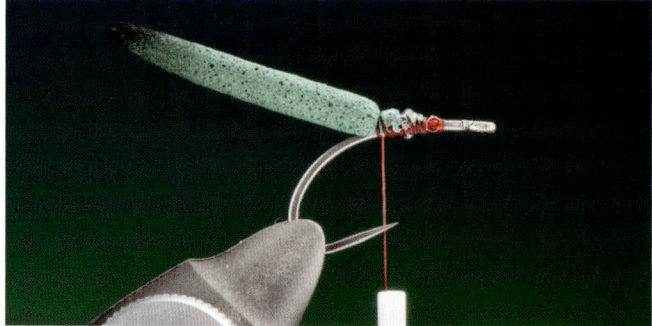

8 Position the prepared extended foam body onto the top of the tying thread foundation and secure with a few wraps of tying thread.

9 If you don't wish to knot your own cock pheasant tail legs, you can buy them pre-knotted as here.

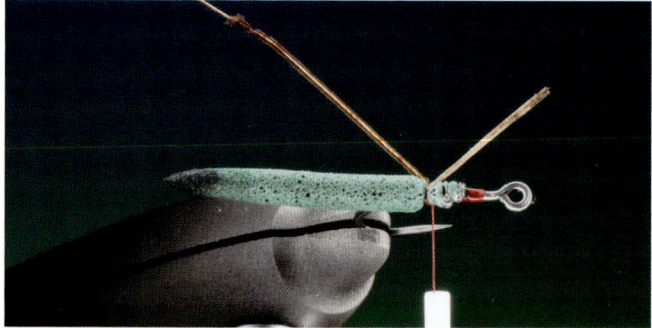

10 If you have a rotary vice, rotate the jaws 45 degrees as shown and tie in the first leg. Take note of the leg length and position in respect to the tying thread foundation.

11 Tie in the second rear trailing leg on the opposite side on the extended body.

12 Select two brown cock hackle tips for the wings. When I am doing demos at large flytying shows, there are often vendors with large boxes of inexpensive Indian cock capes. These are always worth looking through, and every now and then I find a gem!

13 Carefully strip away the lower barbs from the two hackles to form two wings of the same length, a few millimetres longer than the extended body.

14 Position the first wing at the base of the first leg. Secure with only a few wraps of tying thread. If the wing is a little long you can shorten it by gently pulling on the stem.

15 Repeat the last step for the second wing. Both wings should point slightly out, up and rearwards.

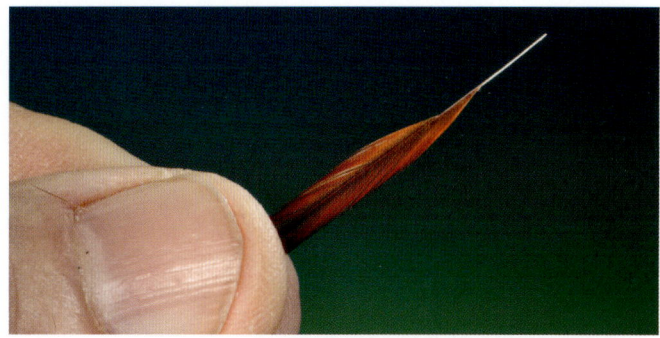

16 You can now prepare a natural brown hackle by stripping off the barbs as shown.

17 Tie in the hackle at the base of the wings.

18 Tie down the hackle stem and then trim away all the surplus material from the legs and wings. Take two more knotted pheasant tail legs and tie these in, central to the thorax as shown.

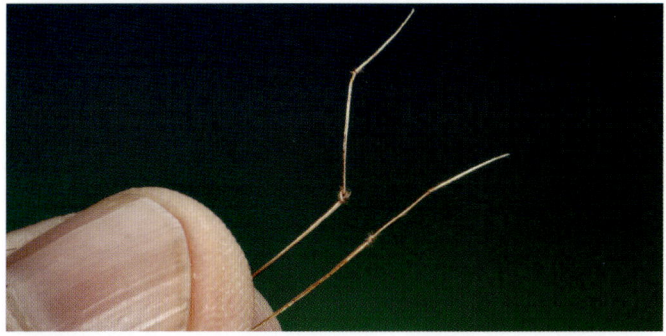

19 You will now need the last two remaining legs.

20 Tie in these two legs on top of the hook shank, facing rearward and up.

225

21 Run your tying thread forward to the hook eye and then fold over the surplus end of the legs and tie these also down rearwards.

22 Trim away the surplus and then fold the front legs forward and tie down into the hook eye.

23 Attach a hackle plier to your hackle and wrap firstly between the rear and central legs, then between the central legs and the front legs and down tight into the hook eye. This should be a densely wound hackle so get in as many turns as possible.

24 Tie off the hackle and remove the surplus, taking care not to crowd the hook eye.

25 While pushing all the hackle fibres and the front legs rearwards, make a couple of whip-finishes.

26 Remove your tying thread and coat the head with a drop of varnish.

27 Side view of the finished crane fly showing leg and wing position along with proportions.

26

Phantom Zonker

Hot glue transparent body
• Mylar overbody • Free-swimming zonker strip wing

The original zonker pattern was tied by the American flytyer Dan Byford in the 1970s. It was quickly recognised the world over as a big fish fly which was extremely easy to tie, yet also a good semi-realistic imitation for most smaller baitfish. In his original pattern, Byford used a lead or tin sheet that was folded and glued over the hook shank and then cut to shape to make the underbody. This hot glue body technique gives the zonker new life. If viewed by a fish in reflected light, the shine and flash of the Mylar body mixed with the animation of the pulsating fur strip makes it a first class baitfish attractor pattern. But when viewed by a fish in a back-lit situation (in silhouette) this pattern really comes to life, with the light penetrating through the transparent hot glue/Mylar body and fur guard hairs.

The flexibility of the Zonker as a baitfish imitation pattern is only as limited as your own imagination. There is a huge amount of rabbit fur strip materials on the market in just about every colour imaginable, not to mention fox, squirrel, mink etc. Then with the vast array of tubing materials available, the combination possibilities are endless for matching the hatch. On this pattern I am using a coypu (*Nutria*) fur strip. Coypu is a very large water-dwelling rodent with unique guard hairs. These are thin at the base of the hair and increase in thickness towards the tips. This gives them extra movement when in the water.

I was first shown this melt glue body technique in 1993 by the innovative Danish flytyer Dennis Jensen who developed it for saltwater seatrout fishing in Denmark. He used a home-made mould constructed from plastic padding. He would insert the hook in the mould and then inject hot glue into it and wait a few seconds for it to dry before removing it. The result was a perfect and identical minnow body every time. Dennis also made very clever subtle body colour changes to his flies by wrapping the hook shank first with tying thread in fluorescent orange, green or blue. Orange was for when he was imitating sticklebacks, green for other small fish and eels and blue when fishing in deep water.

This technique shown here requires no mould. It does take a little practice to master and a few minutes longer, but still produces the same effect.

Another advantage with the zonker, unlike bucktail and featherwing streamers, is that it is an extremely robust pattern. If tied correctly the fly will normally outlive the hook, although the only vulnerability arises if eyes are used and Mylar tubing can be somewhat vulnerable to the small sharp teeth of trout. This can be improved by coating the eyes and Mylar body with UV resin, if wished.

My version of Dennis's original has become one of my most productive patterns for cold water (winter fishing). This whole white pearl, transparent Phantom Zonker has taken many seatrout in the colder months for me. Seatrout, unlike their brown brothers, are seldom selective, especially during the winter months and their menu is 'a la carte' so if whatever you are serving looks and behaves like food, it's on the menu! I have two simple methods of fishing the Phantom. In fairly calm conditions I fish it on a clear intermediate line with a very slow figure-of-eight retrieve, and the takes are normally

fast and brutal. During rough stormy conditions I like to use a floating line and fish the Phantom high in the waves with a short, fast, jerky retrieve, finishing with a slow even lift of the rod. Here the fish will follow creating a bow wave behind the fly and attack on the final lift.

The main advantage of this technique using the transparent hot glue body is that when fishing a streamer, you seldom know where the fish will view it from with regards to the position of the sun! In direct or reflected light, the Mylar body and Ice dubbing will flash and reflect like the scales of a small baitfish. But when viewed in a back-lit situation (silhouette) the body is transparent like that of a small baitfish and not just a solid black silhouette.

TECHNIQUES MASTERED

Hot glue transparent body
- How to sculpt a transparent hot glue baitfish body that works in both reflected and silhouette light situations.

Mylar overbody
- An easy method for using a transparent Mylar tubing over the hot glue body to create a realistic fish scale body effect.

Free-swimming zonker strip wing
- The correct way of attaching a free-swimming fur zonker strip to a streamer hook to reduce hang-ups.

Tying the Phantom Zonker

THE DRESSING

Hook: Partridge Sprite S2200 Barbless Streamer # 6
Underbody: Clear hot glue
Overbody: Pearl Mylar tubing
Thread: Sheer white 14/0
Wing/tail: Fur zonker strip (White Coypu)
Hackle: Pearl Ice dubbing
Eyes: (Optional) tape eyes.

WATCH THE VIDEO

youtube.com/watch?v=_BTUOlfM8pY

Tying the Phantom Zonker with Barry Ord Clarke

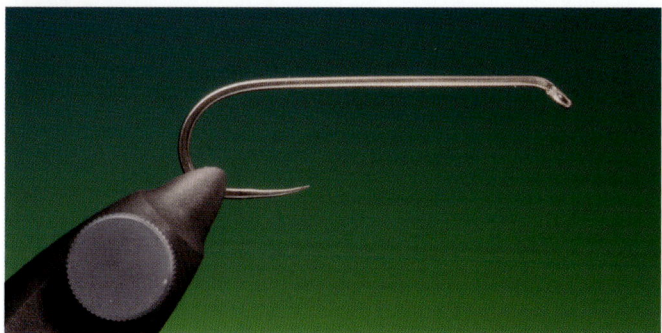

1. Plug in your melt glue gun, as it takes a few minutes to reach temperature. Meanwhile secure your hook in the vice, making sure that the hook shaft is horizontal. If you have a rotary vice, centre the hook.

2. Load your hot glue gun with clear hot glue. Hot glue comes in all manner of colours and even with glitter.

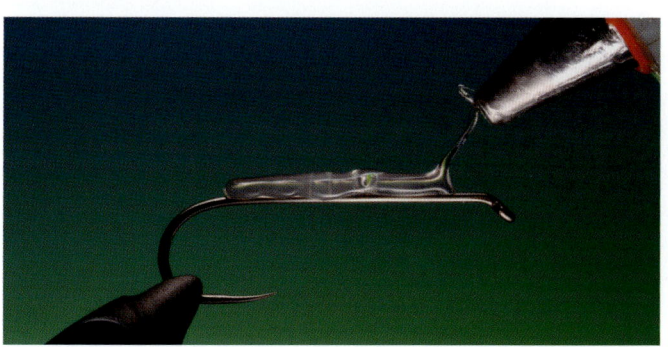

3. When your hot glue gun has reached optimal temperature, run a small amount of clear hot glue along the top of the hook shank as shown. You may find that when you try to remove the melt glue gun, you get a long strand of glue that stretches from the hook to the gun. This can be avoided or resolved by wrapping the strand quickly around the hook shank behind the eye of the hook.

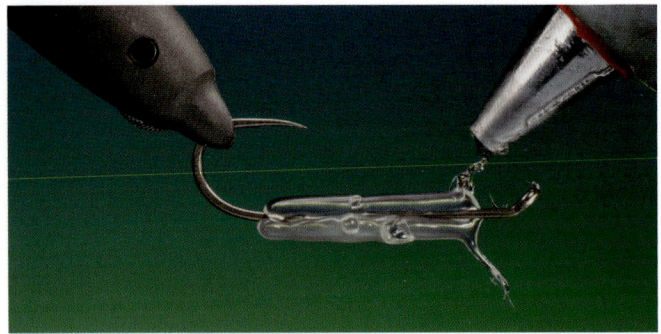

4 Once the glue strip has set a little, about 5 seconds, rotate your vice and carefully apply a little more to form the belly of the minnow. Try and make it the same size as the first strip of glue.

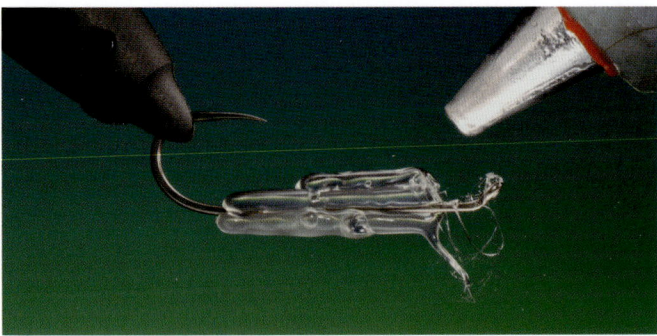

5 After a few seconds, you can apply a smaller amount of glue to the front underside to create the belly. Keep a minnow body shape in mind when applying the glue.

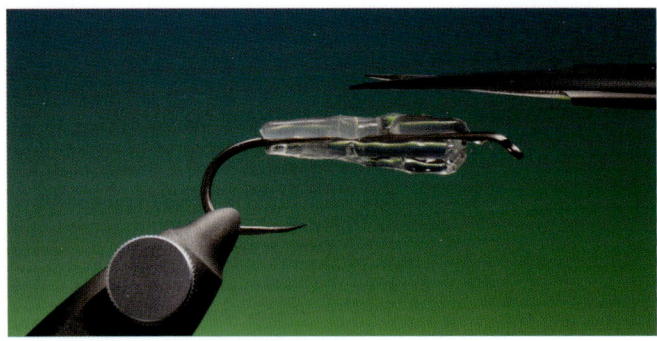

6 Once the glue has set, you must wait a short while, otherwise you will have glue everywhere! You can now trim down the body shape.

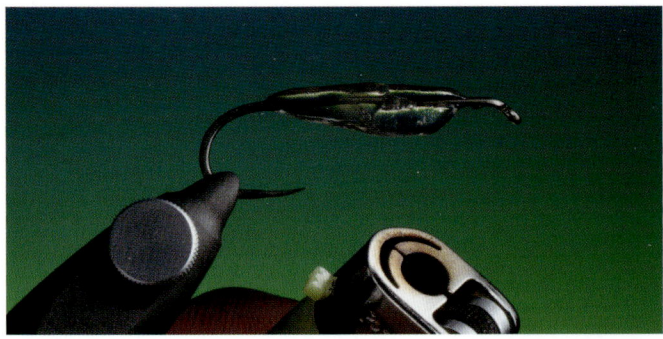

7 You can carefully reheat the body using a cigarette lighter. Take care not to burn the glue, just hold the flame at a distance to warm it up. This will take off the sharp edges.

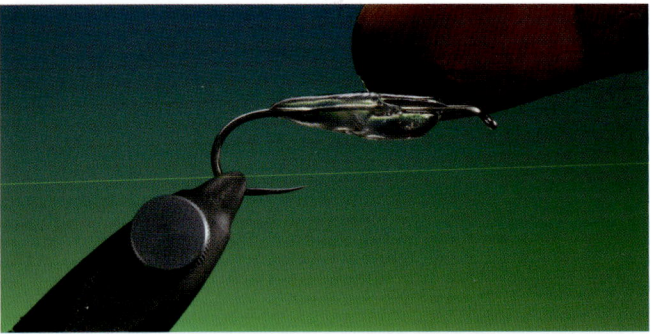

8 When the glue is warm, wet your finger tip. You are then able to shape the body, and this will stop the glue sticking to your finger. It also speeds up the setting process.

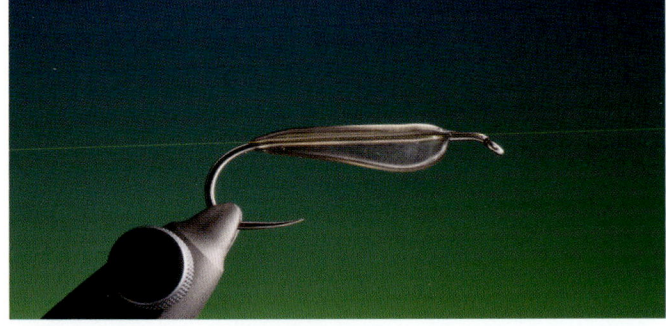

9 Repeat until you are happy with the body shape and size. Then take off the sharp cut edges by warming it again with the lighter. Let your finished body set before continuing.

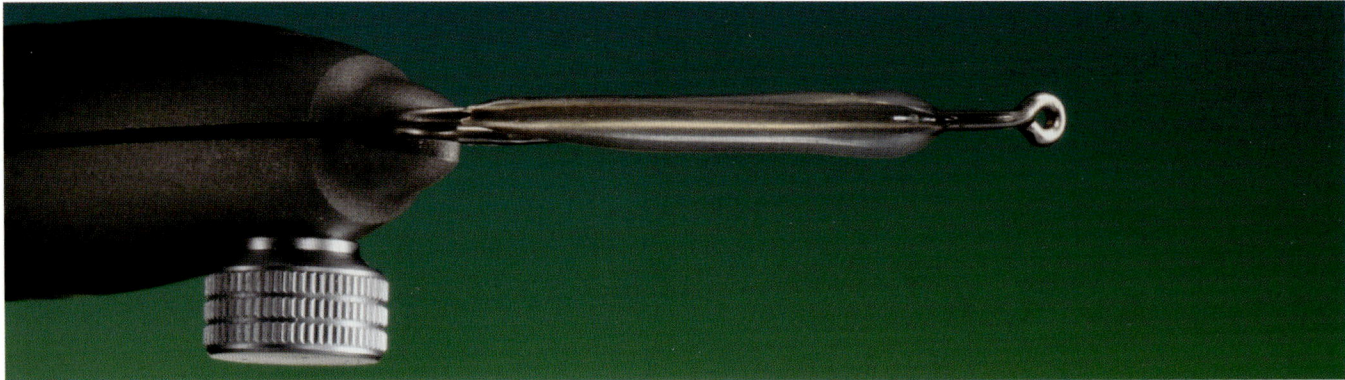

10 Bird's eye view of the slender baitfish body shape.

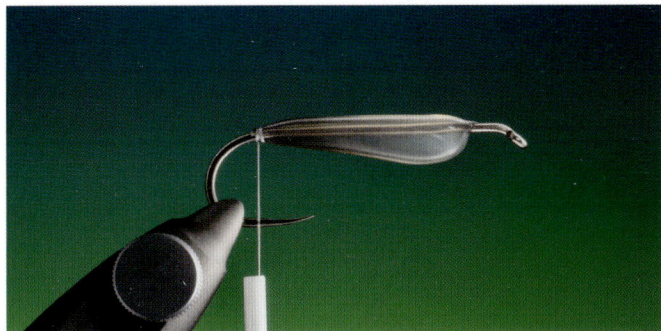

11 Attach your tying thread at the rear of the body.

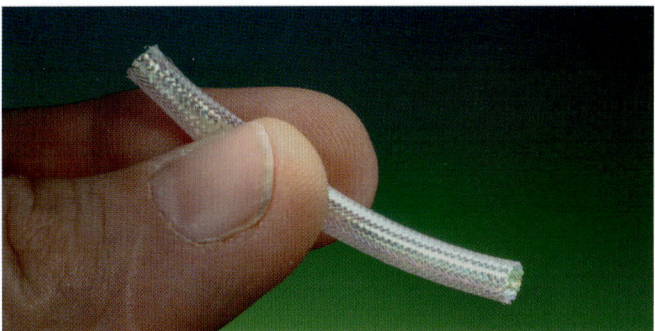

12 Cut a 5-6 cm length of pearl Mylar tubing and remove the string core. Mylar tubing comes in a variety of materials, sizes, diameters, weaves and colours. Not all Mylar tubing works for this particular pattern, so it is advisable to experiment a little beforehand.

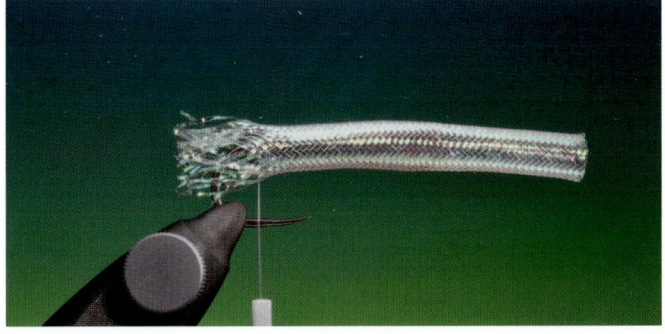

13 Carefully slide the Mylar over the body as shown, with an open weave at the rear. Take care that your tying thread is hanging vertically and not being pushed back by the Mylar.

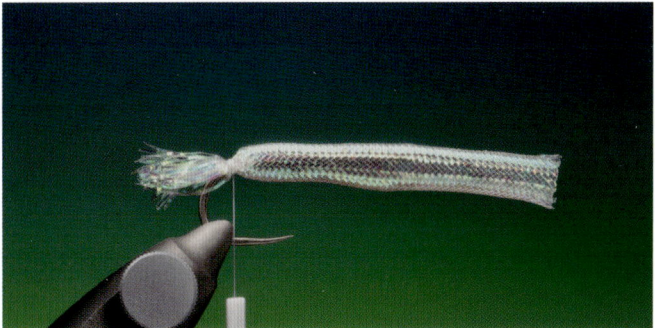

14 Now you can secure the rear of the Mylar with a few wraps of tying thread at the tail base, as shown, leaving a little flashy tail.

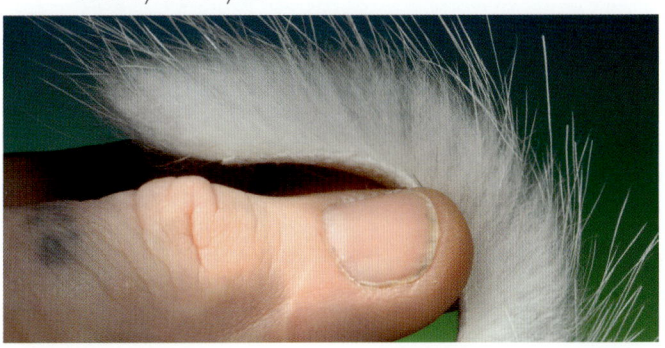

15 Select a length of your chosen fur strip, a little longer than required. Here I am using Coypu (also known as *Nutria*).

16 Before you tie the strip on, prepare the tail end by cutting the hide to a even point. Take care not to cut or damage the fur. This helps the swimming action.

17 Part the fur with the help of a dubbing needle, moist fingers and a clip, at the desired position and then tie it in over the foundation wrappings used to secure the Mylar sleeve as shown. Don't make the tail too long as this will cause it to wrap around the hook bend when fished!

18 Once the zonker strip is in the correct position and secure, give it a couple of whip-finishes and remove your tying thread. Re-attach your thread behind the hook eye, yet tight into the hot glue body.

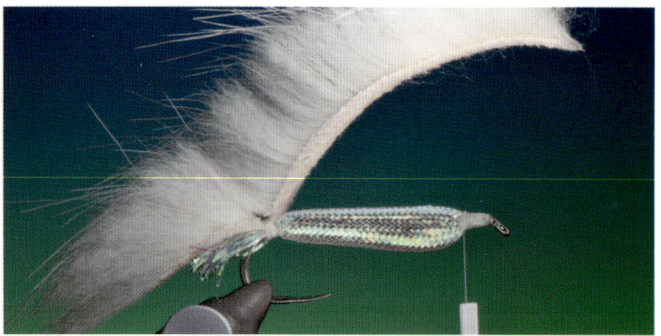

19 Carefully trim away the surplus Mylar tube from around the head, and secure with a few thread wraps.

20 Pull the zonker strip tight over the body. While holding it in position, tie down the strip tight into the hot glue body with a few wraps of tying thread.

21 Carefully trim away the excess zonker strip and build a nice foundation for the Ice dubbing hackle.

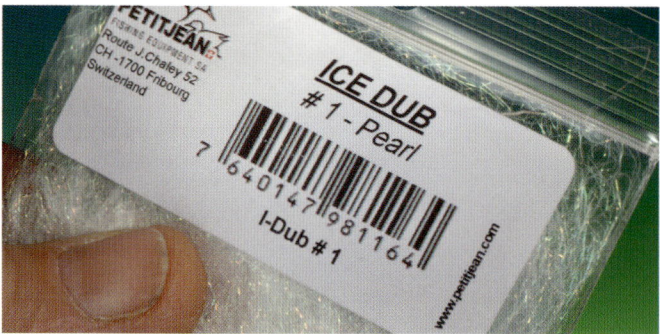

22 You will now need a pinch or two of pearl Ice dubbing.

23 Make a dubbing loop with your tying thread and place a little Ice dubbing sparsely in the loop. Less is more! Spin this up to form an Ice dubbing hackle.

24 Once spun, wrap the Ice dubbing hackle around the head of the fly to form a collar hackle.

25 Then use a dubbing brush or comb to release any trapped fibres in the Ice dubbing, so the collar hackle is evenly distributed around the body and wing of the fly.

26 If wished, you can use a red waterproof felt pen to colour your tying thread before you whip-finish.

27 Whip-finish and remove your tying thread.

28 Finish by giving the head of the zonker a coat or two of varnish.

29 And here's the zonker effect as seen from a back-lit situation, revealing the tiny realistic transparent baitfish body.

30 Here you can see an alternative dressing, with gold Mylar body, squirrel zonker strip and a front hackle. The possibilities are endless!

Gummi Grub Maggot

Lead tape underbody • Clear mono rib • Nymph Skin overbody

Just about every freshwater fisherman has used maggots as bait at one time or another, and that's because they work! Unlike earthworms that get washed into rivers and appear there naturally, maggots are seldom, if ever, found in the trout's diet naturally. But fish love them, and trout and grayling are no exception. They look like food, and are chunky and grubby.

The Nymph Skin is the key to success for these patterns. It makes a very soft, chewy shell for both the maggot and gammarus patterns. Some of the successful colours are orange, pink, olive, white and grey. Nymph Skin comes in all these colours. It couldn't be simpler!

One of the most familiar and readily available amphipods on the fishes' menu is the gammarus, found in large numbers around the shoreline of both fresh and salt waters. The gammarus has a laterally compressed body and is usually seen crawling or jumping about on its side.

There are however a couple of things to pay attention to when tying both these patterns. When wrapping the Nymph Skin, start with it pulled tight, which will make the elastic skin stretch and create less of a build-up on the first few wraps. As you wrap further, release the tension a little as you go, so that as you reach the central position of the body, the Nymph Skin has returned to its original thickness. Using this technique will shape the body.

Also as you are wrapping, take care that each wrap of Nymph Skin covers at least a third of the previous wrap of Nymph Skin. This not only builds the correct body shape again, but it is also the marker for each turn of monofilament rib.

Fishing this pattern couldn't be easier; it fishes shallow without weight or it can fish like a lump of lead weight cutting through water, sinking quickly and fishing deep if required. I use a long fine tippet, in some cases in conjunction with a strike indicator, dead drifting on deep turbulent water. I have two retrieves: a figure-of-eight or a slow short pull.

I have encountered lakes with huge amounts of gammarus and the fish refuse everything else! Every fish I have taken in such lakes were full to the gills with these small freshwater shrimp. Having a good imitative pattern proved to be seriously effective! You will also find that fish that have been feeding on gammarus are normally in exceptionally good condition!

TECHNIQUES MASTERED

Lead tape underbody
- Shaping the underbody and adding weight at the same time with a strip of flat lead tape.

Clear mono rib
- Using a length of clear monofilament as a transparent rib to segment the grub body.

Nymph Skin overbody
- How to attach and stretch Virtual Nymph, Nymph Skin to make a realistic grub/maggot overbody.

Tying the Gummi Grub Maggot

THE DRESSING

Hook: Mustad Heritage C49XSAP # 12-16
Tying thread: Sheer 14/0 White
Weight: Flat lead tape
Underbody: Tying thread
Rib: Clear monofilament
Overbody: Nymph Skin

WATCH THE VIDEO

youtube.com/watch?v=ZIzKxzcb60s

 Tying the Gummi Grub Maggot with Barry Ord Clarke

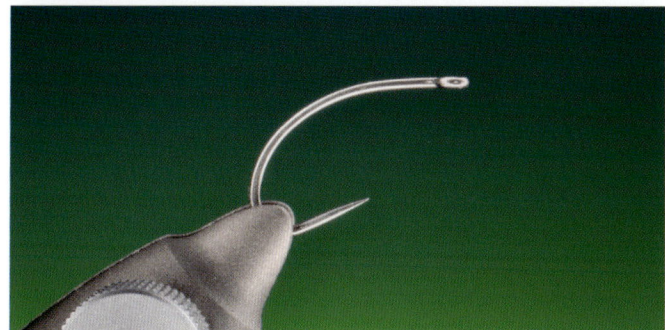

1 Secure your curved grub hook in the vice. Make sure that the hook shaft is horizontal. If you have a true rotary vice, centre the hook.

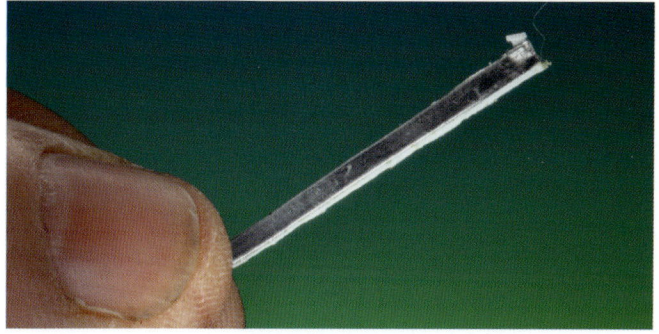

2 From a sheet of lead tape, cut a 2-3mm strip.

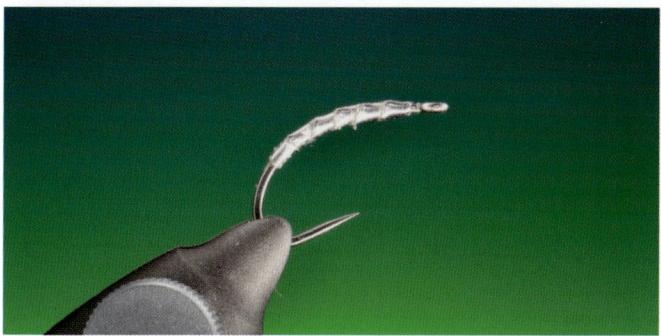

3 Remove the backing from the adhesive tape. Attach the lead strip tape at the rear of the hook and wrap forward in neat touching turns.

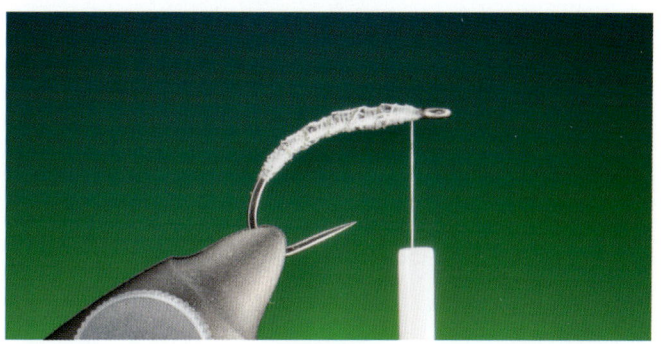

4 Attach your tying thread and spin your bobbin anti-clockwise, until it has a flat profile. Wrap the whole body over the lead tape, to secure it.

239

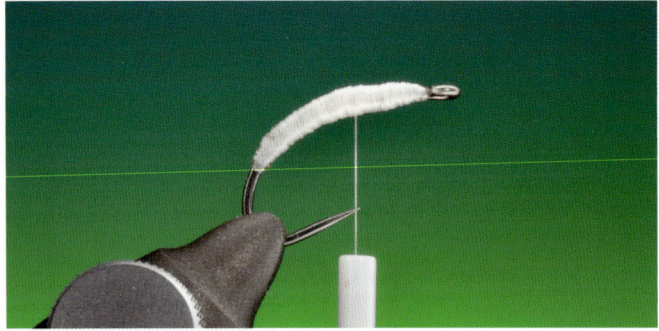

5 Continue wrapping the body with flat tying thread until you have the body size and shape desired.

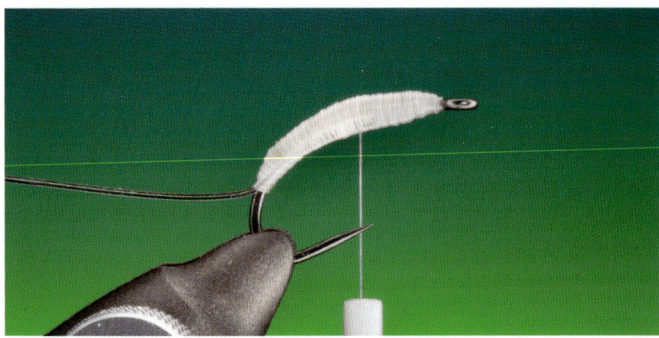

6 Cut a length of clear monofilament for the rib and tie this in along the whole length of the grub body. Finish at the very end of the hook.

7 There are many products on the market that do this job, but the one I prefer is Nymph Skin from Virtual Nymph. Choose your own colour!

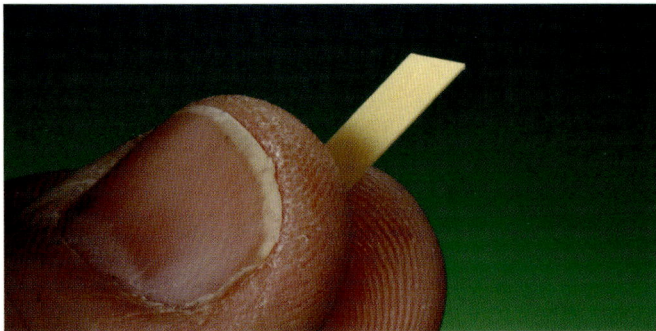

8 To tie this in correctly, once you have cut a length, trim the end at a very slight angle, as shown here.

9 You now need to attach the Nymph Skin at the rear of the hook, by the very end of the cut that you made, and no more. Return your tying thread to the hook eye position.

10 Once secure, you can start to wrap the Nymph Skin. Start by stretching it tightly for the first two wraps and then slacken off as you go, overlapping slightly the previous turn. This will help create the correct body shape. Tie off.

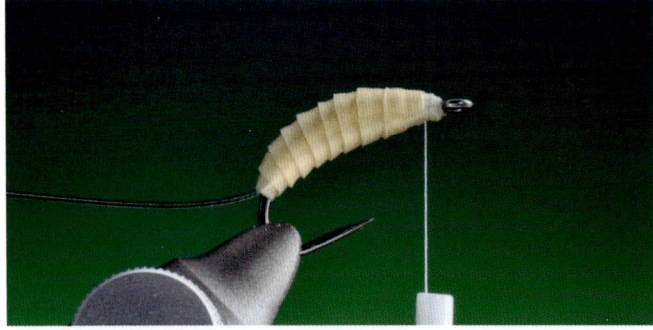

11 If secure, and when you are happy with the body, cut away the surplus Nymph Skin and fix with a few tight wraps of tying thread.

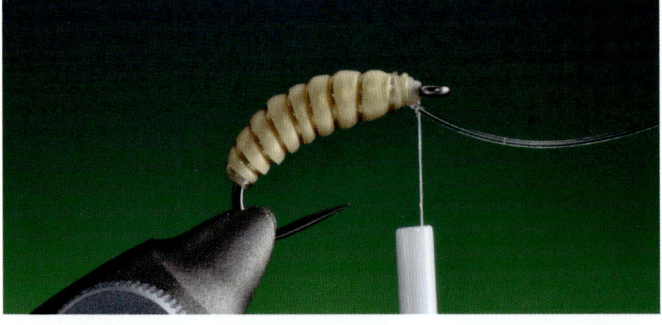

12 Grip the monofilament rib and start to tightly wrap forward. Each turn of rib should fall directly into the rear of each Nymph Skin wrap. Once your reach the hook eye, tie off on the underside of the hook.

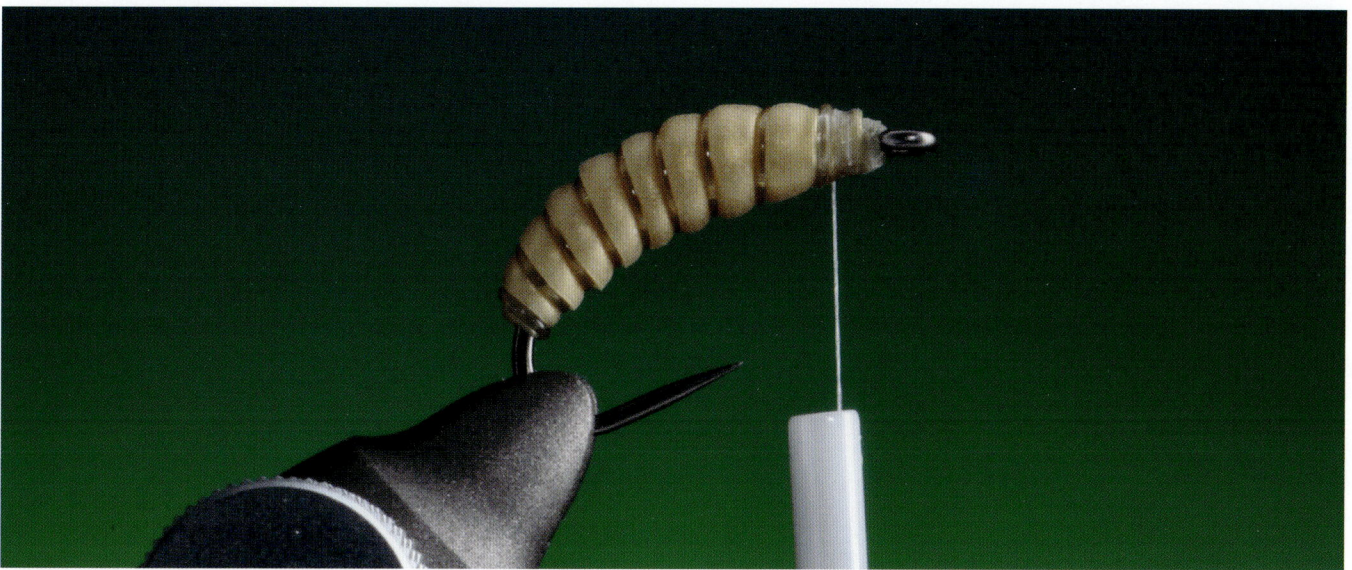

13 Secure the rib correctly, trim away the surplus and secure well.

14 Build up a head with your tying thread. Make a couple of whip-finishes and remove your tying thread.

15 You can use UV resin if you wish for the head, but I find that it doesn't fix well and falls off after a few casts. I like to use Brush Coat Tying Varnish; it's clear, dries quickly and stays on.

16 I give not only the head a coat with Brush Coat but the whole fly.

Alternative Gammarus grub

THE DRESSING

Hook: Mustad Heritage C49XSAP # 12-16

Tying thread: Sheer 14/0 white

Weight: Flat lead tape

Underbody: Tying thread

Rib: Clear monofilament

Legs: Ostrich herl

Overbody: Nymph Skin

17 Remove the backing from the adhesive tape. Attach the lead strip tape at the rear of the hook and wrap forward in neat touching turns.

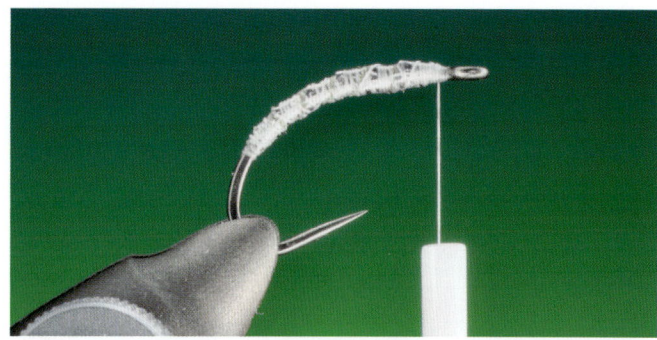

18 Attach your tying thread and spin your bobbin anti-clockwise, until it has a flat profile. Wrap the whole body over the lead tape, to secure it.

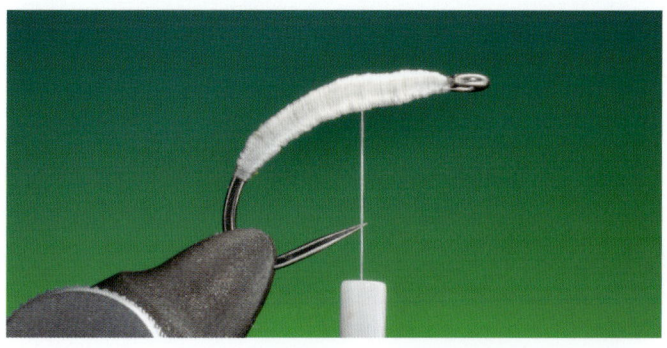

19 Continue wrapping the body with flat tying thread until you have the body size and shape desired.

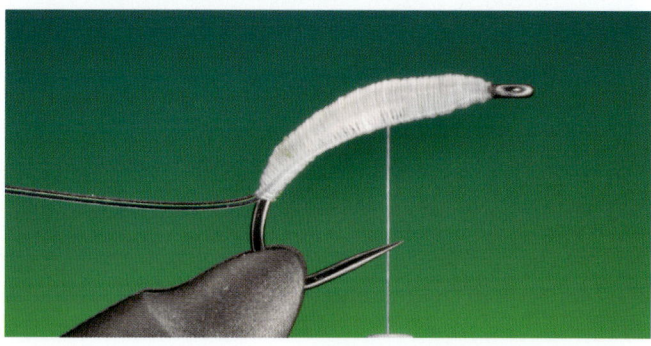

20 Cut a length of clear monofilament for the rib and tie this in along the whole length of the grub body, finishing at the very end of the hook.

21 Take three long ostrich herls.

22 Tie in all three herls at the tail base as shown.

23 You will need a grey Nymph Skin. Attach the Nymph Skin at the rear of the hook, by the very end of the cut that you made and no more.

24 Wrap the Nymph Skin. Start by stretching it tightly for the first two wraps and then slacken off as you go, overlapping slightly the previous turn. This will help create the correct body shape. Tie off.

25 Rotate your vice.

26 Fold over the ostrich herl, and tack down at the hook eye with a couple of loose wraps of tying thread.

27 Rotate your vice again and wrap your rib as before but this time capture the ostrich herl on the underside for the gammarus legs. Tie off the rib.

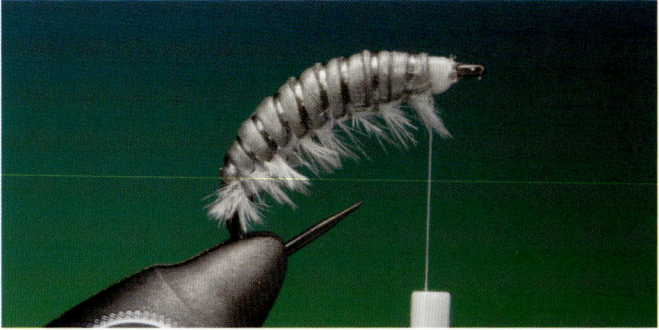

28 Secure the rib correctly, trim away the surplus and secure well.

29 Build up a head with your tying thread. Make a couple of whip-finishes and remove your tying thread.

Silicone Fry

Silicone body • UV resin head • Attractor tape eyes

For many years, I endeavoured to imitate small transparent sand eel elvers, a very popular seasonal, saltwater seatrout food here in Scandinavia. All without any real success, until spring 2014.

In many situations, sea-run brown trout, not unlike their freshwater-based relatives, can be extremely selective when feeding, especially when on these tiny eels, so a good imitative pattern is essential. I tried just about every material, both synthetic and natural, used in other well-known eel and baitfish patterns, but they proved too bulky or too stiff for what I had in mind.

It took a flytying demo in a local tackle shop to solve the problem. After setting up my tying gear ready for the evening demo, with a little time to kill, I had a wander around the shop. While walking through the deep sea fishing aisles of the store, I came across the soft bait section and found the answer to my problem. Muppets, or Silicone Octopus as they are also known.

They are inexpensive, you get loads of material for your money, they come in every colour and size you can imagine, even with glitter and glow-in-the-dark. They were perfect!

This silicone material has other advantages also. It's extremely durable, almost indestructible. It's not too stiff, and not too soft. In fact, it has perfect mobility and swimming action for imitating baitfish and eels.

Albeit a pattern of the utmost simplicity, in both materials and tying skill, it's one of excellent effect. After testing my initial design, a 3-4 cm long transparent elver on a size 14 saltwater hook, I subsequently broadened the pattern to cover other baitfish for both fresh and saltwater.

The eyes on this pattern play no small role. It's worth experimenting with eye colour and size because these are regarded as one of the most important trigger and attack points, for most predatory fish.

Some legs on these soft bait octopus are square not tapered. If so, a better swimming action is achieved if you trim the tail of the extended body to a long fine taper as in step 18. This is best done with a very sharp hobby knife and a metal ruler.

TECHNIQUES MASTERED

A silicone body
- Using a silicone leg from a deep sea octopus soft bait to imitate the extended body of baitfish and eels.

UV resin head
- The careful and gradual application of a UV resin to build a slender baitfish head.

Attractor tape eyes
- Choosing and attaching tape eyes as a predatory fish attractor element.

Tying the Silicone Fry

THE DRESSING

Hook: Mustad Heritage Apex Streamer C81SAP # 8-14

Tying thread: Dyneema or GSP 55

Extended body: Silicone octopus leg

Eyes: Tape or 3D eyes

Head: Clear UV resin

WATCH THE VIDEO

youtube.com/watch?v=Jt4eEAF783M

Tying the Silicone Fry with Barry Ord Clarke

1. Here are just a few colours of silicone octopus available. Each of these has approximately 14 legs. In these sizes you can get two baitfish out of each leg.

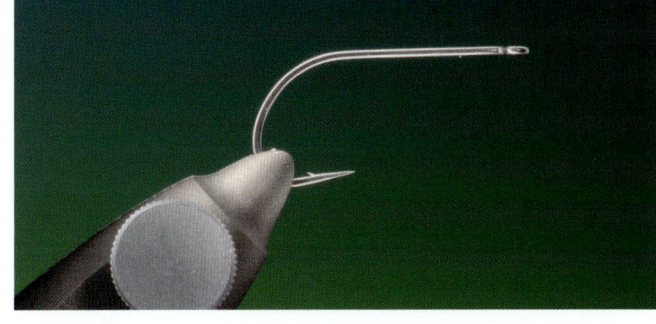

2. Secure your streamer hook in the vice. Make sure that the hook shaft is horizontal. If you have a true rotary vice, centre the hook.

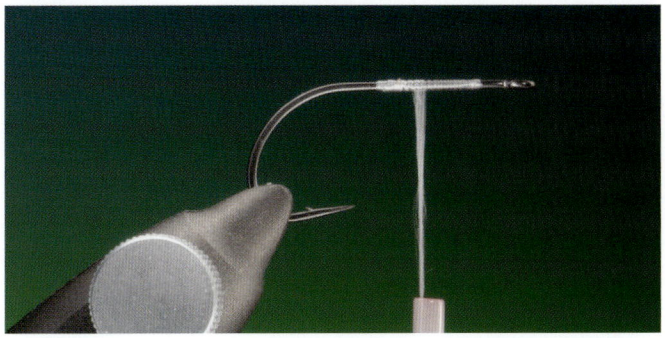

3. Attach your tying thread as shown and cover the front two thirds of the hook shank with a foundation of tying thread.

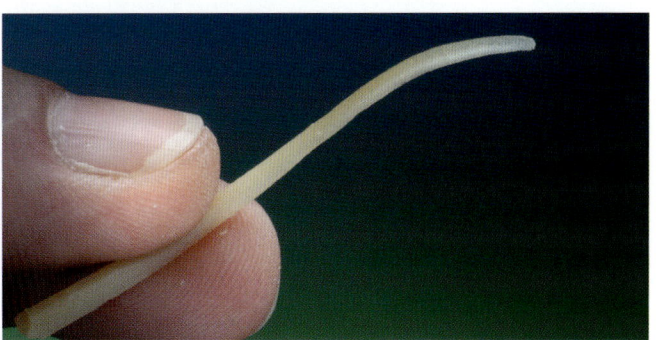

4. Select the colour of silicone octopus leg you require and cut to the desired length.

5 Spin your bobbin anti-clockwise so that it attains a flat profile. This will prevent the tying thread from cutting through the silicone. Only attach the silicone the same distance as the tying thread foundation. This will give a better swimming action.

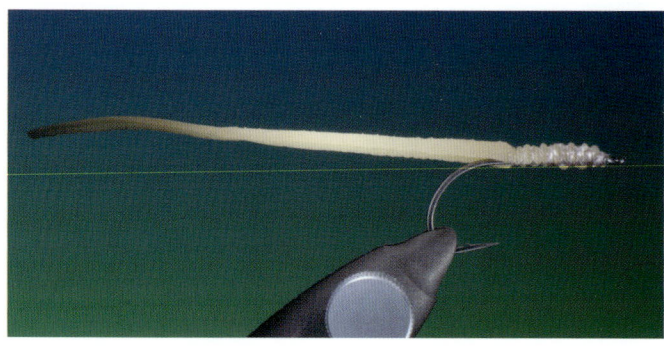

6 Take care that the silicone extended body is central and balanced on the hook shank. Tie off with a couple of whip-finishes and remove your tying thread.

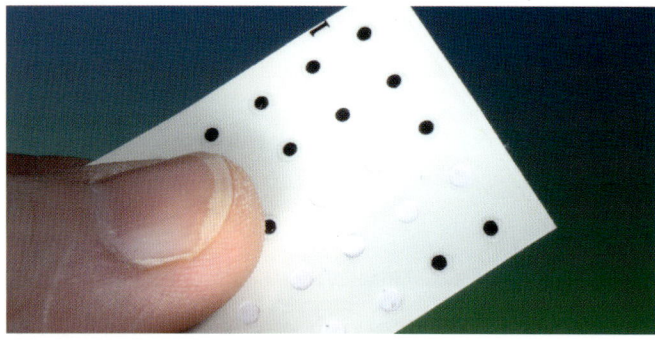

7 You will need some eyes. For the smaller patterns I prefer to use small tape eyes. These are an essential attack stimulator for most predatory fish.

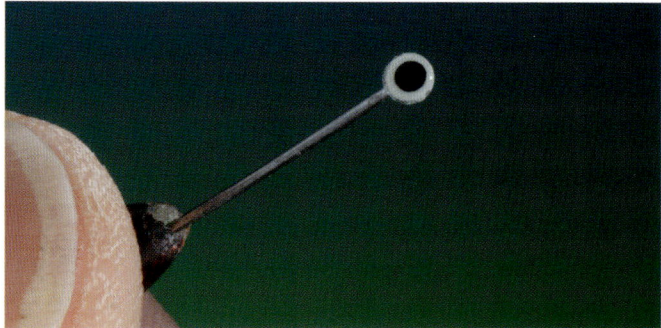

8 I find that the easiest method of removing the eyes from the sheet and attaching them is by using the point of a dubbing needle.

9 Place one eye each side of the head, taking care again that they are balanced. The tape on these eyes is occasionally not adhesive and if so, use a little superglue to hold them in position.

10 Here you can see the that the eyes are balanced on each side of the baitfish head and the body is balanced.

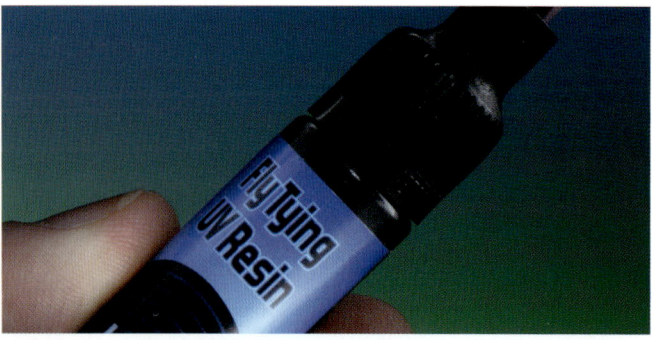

11 You will need a clear UV resin to secure and build up the head.

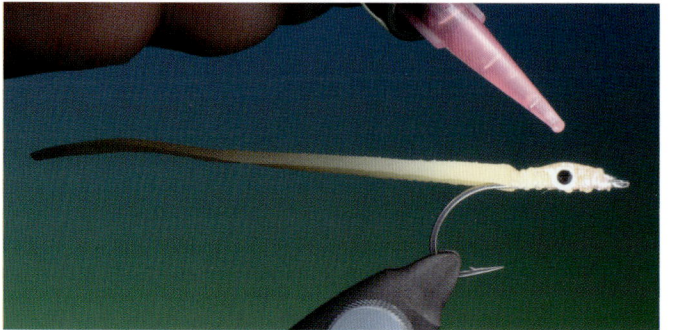

12 When applying the UV resin, do so only a little at a time, then cure with the light. This takes a little longer but reduces the chances of the UV resin running and making mistakes.

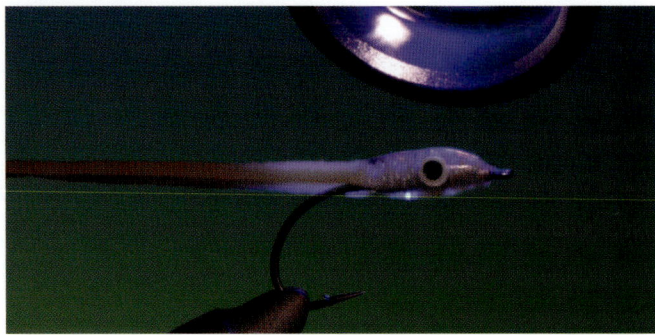

13 Once you have built up the initial head you can finish it with a final overall coat of UV resin and a longer cure with the torch.

14 The finished head should be even and small.

15 Bird's eye view of the slender head and body of the streamline baitfish.

16 This is the original sand eel elver or glass eel as it is also called. Note the red gills. This is another attack point, easily achieved with a red felt pen, then given a coat of UV resin over.

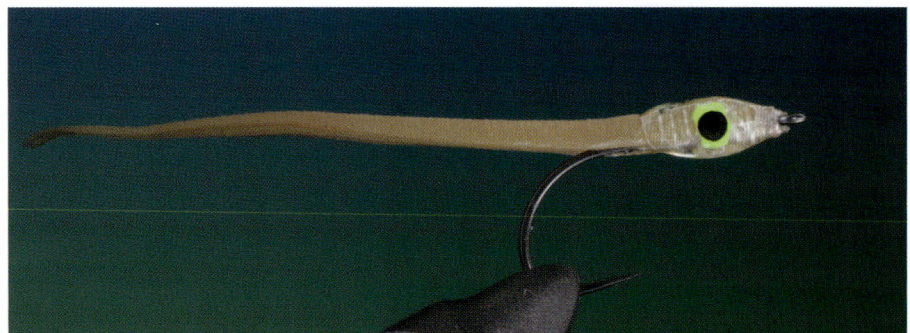

17 Here's a much smaller version with bright yellow eyes as an added attractor.

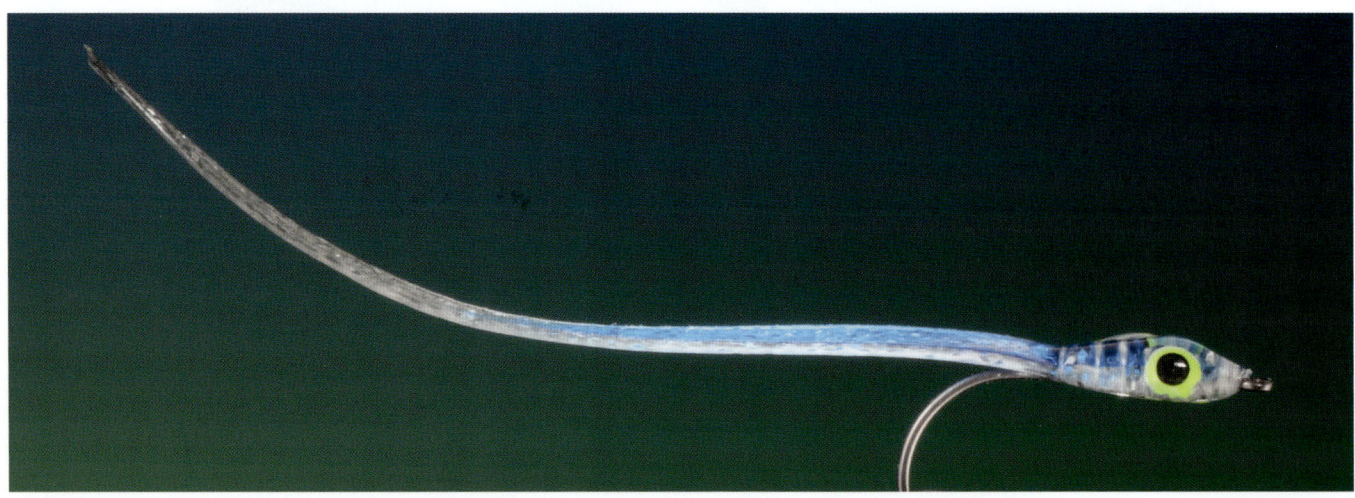

18 Many of the silicone octopus legs have a gradated colouration. This one starts with blue and tapers into clear with glitter in both. I have found this very effective as a sand eel imitation.

THE AUTHOR

Born in England, Barry Ord Clarke is an internationally acclaimed flytyer, photographer and author. He has won medals in some of the world's most prestigious flytying competitions, and his own flies can be seen in the iconic Flyfishers' Club collection in London and in the Catskill Master Fly Collection in the Catskill Museum in the United States.

In 2016, he was awarded the coveted Claudio D'Angelo award for Best International Fly Tyer.

In 2021 Barry was voted *Fly Tyer magazine*'s 'Fly Tyer of the Year'. Barry was the first person to receive this prestigious award. It honours the exceptional international contribution Barry has made to the world of flytying with his innovative approach of linking his clear step-by-step tying instructions in book form to his popular YouTube videos which then show him tying EXACTLY the same fly online.

For the past 30 years he has lived in Norway where he works as a professional photographer and fly tying consultant for Mustad, and Veniard Ltd.

You can find Barry's flytying demonstrations on his successful blog and YouTube channel, The Feather Bender. www.thefeatherbender.com

Useful websites & organisations

54 Dean Street
www.54deanstreet.com
info@54deanstreet.com

Ahrex hooks
www.ahrexhooks.com
info@ahrexhooks.com

C&F Design
www.c-and-f.co.jp

Chevron Hackles
www.chevronhackles.com
Chevronhackles@yahoo.co.uk

Cookshill Flytying Materials
www.cookshill-flytying.co.uk

Farlows
www.farlows.co.uk

Fly Box Direct
www.flyboxdirect.co.uk

Fly Fish Europe
www.Flyfisheurope.com

Fly Only Online
www.flyonlyonline.co.uk

Fly Shack
www.flyshack.co.uk

Fulling Mill
www.fullingmill.co.uk

Gallo de León
Javier Escanciano
www.gallodeleon.com

Glasgow Angling Centre
www.fishingmegastore.com
sales@fishingmegastore.com

Hareline Dubbin Inc
www.hareline.com
hareline@hareline.com

John Norris of Penrith Ltd
www.johnnorris.co.uk

Lakeland Flytying
www.lakelandflytying.com

Live 4 fly fishing
http://live4flyfishing.pl/
live4flyfishing@gmail.com
info@live4flyfishing.pl

Mustad Hooks
www.mustad-fishing.com

MvE Products
www.maartenvaneijk.nl

Orvis
www.orvis.co.uk (UK)
www.orvis.com (USA)

Partridge of Redditch
www.partridge-of-redditch.co.uk

Peak Fishing
Peakfishing.com
info@peakfishing.com

Petitjean Fishing Equipment
www.petitjean.com
info@petitjean.com

Polish Quills
www.polishquills.com
info@polishquills.com

Rainy's Flies
www.rainysflies.com

Renomed
www.renomed.eu/?lang=en

Semperfli
www.semperfli.net

Sportfish
www.sportfish.co.uk

The Essential Fly
www.theessentialfly.com

The Flydressers' Guild
www.flydressersguild.org

The Fly Fishing Show
www.flyfishingshow.com

The Fly Tying Company
www.flytyingcompany.co.uk

Trout Catchers
www.troutcatchers.co.uk

Trout Line
www.troutline.ro
office@troutline.ro

Veniard Ltd
www.veniard.com
sales@veniard.com

Virtual Nymph Products
www.virtual-nymph.com
sales@virtual-nymph.com

Wapsi Fly
www.wapsifly.com
pat@wapsifly.com

Whitetail Fly Tieing Supplies
www.whitetailflytieing.com

INDEX

A

All Fur Wet Fly	**79–85**
Anorexic Mayfly	**95–101**
Antron yarn overbody	63, 64
ants	205
attractor tape eyes	245, 246

B

Bingen, Gunnar	141, 142
bouyancy & flies	17–18
Braided Blue Damsel	**185–192**
braided extended body	185
Byford, Dan	229–236

C

Catskill Museum	251
CDC	14, 17, 57, 87, 88, 95, 103–109, 125, 175
CDC breathing gills and wing	57–62
CDC dubbing brush thorax	175, 176
CDC dubbing brush wing and hackle	125, 126
CDC extended body	95, 96
CDC Mayfly Dun	**103–109**
CDC parachute wing	95, 96
CDC para post	87, 88
CDC split wings	103, 104
CDC wings	175, 176

Clarke's All-Purpose Emerger	**51–56**
Clarke's Caddis	**125–132**
classic dubbed mayfly body	119, 120
clear mono rib	237, 238
Comparadun	95, 96, 97, 100, **119–124**
Coq de Leon comparadun wing	95
crane fly	221, 223

D

daddy long-legs	221
D'Angelo, Claudio	251
Danica Mayfly Nymph	**35–41**
deer hair bullet head and wing	213, 214
deer hair Comparadun wing	119, 120
deer hair hackle	51, 52
deer hair leg construction	205
deer hair shell back	51, 52
deer hair tail	119, 120, 141, 142
deer hair tail and head in one	141, 142
deer hair wing	63, 64
Diving Caddis	**133–140**
double hackle tip wings	167, 168
double rubber legs	213, 214
double Wally wings	157, 158
dubbed dry fly body	141, 142
Dyret	**141–148**

253

E

Eijk, Maarten van	175
elk hair down-wing	125, 126
elk hair extended body	125, 126
Emergent Sparkle Pupa	**63–70**

F

Feather Bender	3, 7, 8, 251, 256
flat copper wire	37, 57, 59, 58, 59
Float Foam Ant	**205–212**
floss silk body	213, 214
Fluttering Caddis	**157–166**
Fly Tyer of the Year	251
foam body	205, 206
Foam Cylinder Crane Fly	**221–228**
foam extended bodies	221, 222
foam para post	185
free-swimming zonker strip wing	229, 230
fur hackle	79, 80, 81

G

Giant Stone Fly	**175–184**
Golden pheasant tippet tail	167, 168
Gummi Grub Maggot	**237–244**

H

hackle stem antennae	149, 150
hackle tip wings	205, 206, 221, 222
Hare's ear body	71
Hare's Ear dubbing technique	87
Hare's Ear Parachute	**87–93**
Hare's Ear Soft Hackle	**71–77**
heavy front shank head	193, 194
hooks, background	10-11
hot glue transparent body	229, 230

I

Ice dubbing thorax	185, 186

K

Klinkhåmer	7, 11, 14, 28, 32
Klinken, Hans van	7

knotted pheasant tail legs	221, 222
Krystal Flash trailing shuck	51, 52, 54

L

LaFontaine, Gary	63
lead tape underbody	237, 238
lead wire flattening technique	79, 80

M

Madam X	**213–220**
mallard feather caddis wings	149, 150
mallard flank antenna	157, 158
Mallard Slip Wings	**111–118**
mallard wing	113, 133, 134
Midge Emerger	**57–62**
moose body hair tail	103, 104
Moose Mane Nymph	**43–49**
Muppets	245
Mylar overbody	229

N

Neon hot spot	133, 134
nymph legs	28, 39, 43, 44
Nymph Skin overbody	237

P

palmered body hackle	141, 142, 167, 168
Panama Palmer	167–174
parachute hackle	14, 87, 88
partridge body hackle	133, 134
Phantom Zonker	**229–236**
proportions of flies	13-15

Q

QR code	8
quill slip wings	111, 112
quill-style body	43, 44

R

rear weighted wire ribbing	35, 36
Red Panama	**167–174**
rubber legs	213, 214, 217, 218

S

segmented abdomen technique	157, 158
silicone body	245
Silicone Fry	**245–250**
Silicone Octopus	245
slip wings	111, 112
soft swimming body	193, 194
speckled partridge front hackle	167
spinning fur and Ice dub	193, 194
split mayfly tails	27, 28
split pheasant tail legs	27, 28
Stone Fly bodies	175, 176
stripped hackle nymph legs	43, 44
Swisher, Doug	213

T

tapered fur body	79, 80
The Animal	141, 143
traditional style dry fly	111
trailing shuck	53, 63, 64, 65
twisted marabou dubbing technique	35, 36
twist & wrap CDC body	149
two tone wing case	43, 44
tying thread rib	103, 104

U

Ubiquitous Nymph	8, **27–33**
UV resin head	245, 246

V

varnishing heads	22–23

W

Wally wings	157, 158
wax	18–20
Welshman's Button	**149–156**
wet fly soft hackle	71, 72
wing cases	27, 28
wood duck tail	71, 72
Worm	**193–204**

Y

YouTube	7, 8 19, 95, 251

TYING NOTES

Also published by Merlin Unwin Books

The Feather Bender's Flytying Techniques - Barry Ord Clarke
Flytying for Beginners - Barry Ord Clarke
The Klink - Hans van Klinken
Pocket Guide to Matching the Hatch - Peter Lapsley & Cyril Bennett
Pocket Guide to Fishing Knots - Peter Owen
Tying Flies with CDC - Leon Links